RAND NATIONAL SECURITY RESEARCH DIVISION

Enhancing Next-Generation Diplomacy Through Best Practices in Lessons Learned

Dwayne M. Butler, Angelena Bohman, Christina Bartol Burnett, Julia A. Thompson, Amanda Kadlec, Larry Hanauer

Sponsored by the Una Chapman Cox Foundation

For more information on this publication, visit www.rand.org/t/RR1930

Library of Congress Cataloging-in-Publication Data is available for this publication.
ISBN: 978-0-8330-9829-0

Published by the RAND Corporation, Santa Monica, Calif.
© Copyright 2017 RAND Corporation
RAND® is a registered trademark.

Support RAND
Make a tax-deductible charitable contribution at
www.rand.org/giving/contribute

www.rand.org

Preface

Reflective and adaptive organizations are effective organizations. These characteristics are particularly important in today's complex world. The 2015 Quadrennial Diplomacy and Development Review acknowledged that "today's international landscape is more complex than ever before" and lists three lines of effort to promote the innovation necessary to meet these challenges: "support creative problem-solving," "institutionalize policy to encourage innovation while managing risk," and "capture and communicate lessons learned."[1]

This research provides theoretical backdrop and exploration into best practices across various fields to prescribe how the Department of State could continue to develop a strong culture of learning and implement an enterprisewide lessons-learned capability. This report should be of interest to organizations seeking to establish, conduct, and sustain effective lessons-learned activities.

The research was conducted within the International Security and Defense Policy Center of the RAND National Security Research Division (NSRD) under a contract for services with the Una Chapman Cox Foundation. NSRD conducts research and analysis on defense and national security topics for the U.S. and allied defense, foreign policy, homeland security, and intelligence communities and foundations and other nongovernmental organizations that support defense and national security analysis.

For more information on the International Security and Defense Policy Center, see www.rand.org/nsrd/ndri/centers/isdp or contact the director (contact information is provided on the web page).

The opinions and characterizations in this report are those of the authors and do not necessarily represent official positions of the U.S. government.

---

[1] U.S. Department of State, *Ensuring Leadership in a Dynamic World: Quadrennial Diplomacy and Development Review*, Washington, D.C., 2015, pp. 57–58.

Contents

Figures and Tables

Figures

Tables

Summary

Background

Reflective and adaptive organizations are effective organizations. These characteristics are particularly important in today's complex world. The 2015 Quadrennial Diplomacy and Development Review acknowledged that "today's international landscape is more complex than ever before" and listed three lines of effort to promote the innovation necessary to meet these challenges: "support creative problem-solving," "institutionalize policy to encourage innovation while managing risk," and "capture and communicate lessons learned."[1]

The creation of the Center for the Study of the Conduct of Diplomacy in 2014 shows that the Department of State is committed to becoming more reflective and adaptive. As part of the Foreign Service Institute (FSI), the center is responsible for "disseminat[ing] case studies, includ[ing] them in training exercises, and integrat[ing] them into interagency and community-wide planning."[2] The center currently examines the practice of diplomacy, undertaking policy implementation reviews, which compare how a single issue area was tackled in several different cases. The current primary use for these reviews is in FSI training courses, where the lessons learned are used to improve instruction materials and approaches.

In addition to the Center for the Study of the Conduct of Diplomacy, other existing Department of State organizational elements and attributes are active in many aspects of the lessons-learned process and provide a starting point for expanding and integrating the department's current lessons-learned capabilities. An enterprise plan that bolsters and enhances the effectiveness of these capabilities will help organizational units maintain the cultural acceptance necessary for an effective program that they have gained so far. To that end, this research provides a theoretical backdrop and explores best practices across fields of organizational, educational, and learning theory to help the Department of State consider how it might further develop its lessons-learned practices.

---

[1] U.S. Department of State, *Ensuring Leadership in a Dynamic World: Quadrennial Diplomacy and Development Review*, Washington, D.C., 2015, pp. 57–58.

[2] U.S. Department of State, 2015, p. 58.

This report does not assess or otherwise comprehensively measure the Department of State's current lessons-learned capabilities and does not prescribe a tactical implementation plan. For contextual reasons, it does generally discuss departmental information gleaned from our efforts to familiarize ourselves with the Department of State and from interactions with sections of the organization. Moreover, our use of general contextual information sets conditions for a presentation of thematic best practices for next steps in the evolution of the department's enterprise lessons-learned initiatives.

Research Questions and Approach

The following questions guided our research and highlight the Department of State's key interests regarding best practices in lessons learned:

- How do organizations undertake lessons-learned efforts? How do they ensure that such efforts are comprehensive, fact-based, and free from undue bias or influence? What research and evaluation techniques are effective and ineffective? What organizational structures make such efforts most effective?
- How are lessons-learned efforts managed and overseen in such a way that employees see them as organizational priorities and as performance-enhancing exercises instead of as efforts to assign blame for mistakes?
- How do organizations use the results of lessons-learned studies to change practices and behavior? How are lessons incorporated into professional training?
- How, if at all, do efforts to study and apply lessons learned differ in the public and private sectors?

To identify best practices in lessons learned and to assess how organizations use lessons-learned initiatives to promote a culture of learning, we reviewed the literature and collected qualitative data on three primary categories: military, other government organizations, and the private sector. To analyze the data collected within the research categories, we developed a list of research lenses relevant for the Department of State's work: purposes of lessons learned; vision; enabling elements; collection planning; approaches to collection, validation, and dissemination; impact; knowledge management; talent management; communities of practice; and overcoming obstacles. We organized insights in each research lens into one of three groups: observation, best business practices, and recommendations.

We also present two evaluation tools that the Department of State and other organizations can use to examine the growth of their lessons-learned programs and to guide planners to consider a wide range of issues: the Maturity Implementation Model an interviewee offered us and a framework we developed over the course of the study, the Lessons-Learned Capability Assessment Framework. We provide contextual background on some of the Department of State's initiatives and apply this context to the Maturity Implementation Model. It identifies key stages of a program's growth and maturity. To guide planners as they develop next steps, we created a second framework derived from our research, the literature, and expert opinion. This model, which we refer to as the Lessons-Learned Capability Assessment Framework, examines formalization (whether an office is mandated by law or suggested by an individual leader), governance (how much authority that office has to affect change), human capital management (the hiring, management, and growth of personnel with competencies necessary to conduct successful lessons-learned activities), and resources.

Summary of Findings

We found that best practices differ, often greatly, depending on which strategic vision or approach an organization selects. We found that vision and approach matter far more than whether an organization is private or public. Examples throughout this report highlight this dynamic and address successful techniques for conducting and managing a lessons-learned program. Other findings include the following:

- Designating an integrating or steering element (office, department, other group) helps organizations drive lessons-learned efforts.
- Organizations can use three primary strategic approaches to drive lessons-learned programs and/or the transfer of knowledge:
  - business process management (structured learning executed at all levels of an organization, led by managers as part of their day-to-day responsibilities)
  - corporate learning strategies (a separate institution within an organization that helps the entire organization acquire, apply, and share knowledge[3])
  - hybrid of both.
- Developing a vision for the lessons-learned program is key to its success. The vision is necessary to answer essential implementation decisions:
  - What are the desired outcomes of the program?
  - What approaches will be used to achieve these outcomes (business process management and/or corporate learning strategies)?
  - Whom do you wish to reach in the organization (individuals through leadership and/or business units)?
  - How will the lessons-learned program be governed (centralized or decentralized), given desired outcomes and target audience?

---

[3] International Institute for Management Development, "Corporate Learning," Insights@IMD, No. 4, Lausanne, Switzerland, May 2011.

- Many approaches to collection, validation, and dissemination exist. Deliberate choices need to be made on these matters and in general to clearly guide the growth of lessons learned.
- A strong culture of learning is essential for a lessons-learned program to function effectively and can be fostered through *intra*organizational enabling elements, such as securing leader and workforce buy-in; establishing practices, policies, and procedures; and creating and/or connecting existing organizational infrastructure.
- *Extra*organizational knowledge sources, including communities of practice and informal networking, can enhance the effectiveness of the lessons-learned program.
- Many challenges to creating a culture of learning exist, including reticence to sharing failures, resource constraints (personnel and/or funding), and the need to demonstrate added value or impact.
- *Intra*organizational elements, coupled with open and inclusive collection methods, can help leaders hedge against undue biases or influence.
- Knowledge management—making collected information and lessons that have been learned available to the broader organization—is a major aspect of the lessons-learned effort.

Thematic Best Practices

We provide the following thematic best practices to guide the Department of State's next steps in developing its lessons-learned program:

1. *Develop an annual collection plan that flows from the organization's vision and mission and get senior leadership buy-in on what is to be collected and disseminated.* This approach helps determine the offerings of the lessons-learned capability and sets conditions for staffing and resourcing decisions for the lessons-learned program. Collection planning should include determining triggers or learnable events and when to collect on these events. Collection could occur at all stages of an employee's assignments or career progression, for example, at the end of FSI training, in the middle of an embassy tour (ad hoc), and on returning from an assignment. This would allow Department of State employees to be consumers during training and consumers and producers once in the field.

2. *To establish a tangible list of requirements, determine the specific outcomes lessons-learned initiatives should achieve.* In Chapter Two, this report provides several examples of different categories of goals, including institutional knowledge and risk avoidance. Once the requirements for achieving the desired outcomes have been established, the department should identify the current centralized and

decentralized and formal and informal capabilities currently in place to determine where additional capability needs to be created and where existing capability can be integrated into an enterprisewide program.

3. *To better govern the lessons-learned program, assign an office or body primary responsibility.* Clear leadership in this sense can help the organization remain focused on the program's vision and goals. Our research has shown that organizations use formalized products, such as policies, performance reviews, and incentive programs, to drive the culture of learning. The Department of State should consider implementing a combination of these formal tools to shape its learning culture.

4. *Establish a formal, named campaign with an accompanying slogan to promote the culture of learning and the existence of the formalized lessons-learned capability.* The organizations at which we conducted interviews—such as the Central Intelligence Agency's Center for the Study of Intelligence—developed robust marketing campaigns that include posters physically hung around their buildings.

5. *Develop a resource-allocation strategy to ensure that resources are not a constraint to the lessons-learned program's ability to meet the desired outcomes of the program.* Begin with personnel analysis to determine the number and type of staff required and then use innovative ways to insource and outsource those requirements.

Acknowledgments

The authors would like to thank the Una Chapman Cox Foundation for funding this study and the Department of State's Center for the Study of the Conduct of Diplomacy and its parent organization, the Foreign Service Institute, for their guidance and support. The authors would also like to thank all our interviewees for their participation in this study and their involvement in the broader community of practice. Finally, we would like to thank Charles Ries for his guidance over the course of the study and our reviewers, James Dobbins and James Bailey, for their helpful commentary.

CHAPTER ONE

Introduction

In recognition of the need to modernize, many government agencies have started implementing strategies to motivate organizational change and improvement. The U.S. Department of State is one such agency. Its mission is to "shape and sustain a peaceful, prosperous, just, and democratic world, and foster conditions for stability and progress for the benefit of the American people and people everywhere."[1] In pursuit of this mission, one of the five major strategic goals for the Department of State is to "modernize the way we do diplomacy and development."[2]

Targeting this goal specifically, the 2015 Quadrennial Diplomacy and Development Review (QDDR) acknowledged that "today's international landscape is more complex than ever before," and listed three lines of effort to promote the innovation necessary to operate in this complex environment: "support creative problem-solving," "institutionalize policy to encourage innovation while managing risk," and "capture and communicate lessons learned."[3] It further stressed that the modernization strategies for the next generation of diplomats should provide "both the willpower and the tools to accomplish the impossible," which, in the context of this report, is the tradecraft of diplomacy.

The 2015 QDDR established the framework for implementing lessons-learned efforts throughout the Department of State—that is, efforts to learn from experience, including both successes and failures, and to share those lessons throughout the department to achieve organizational change. The Department of State has made significant strides in establishing an enterprisewide lessons-learned capability and wishes to mature and expand its efforts in the future.

The Center for the Study of the Conduct of Diplomacy (CSCD) of the Foreign Service Institute (FSI) serves as an example of one of the Department of State's centralized, formal structures to promote a culture of learning by providing tools and

---

[1] U.S. Department of State, *Strategic Plan FY 2014–2017*, Washington, D.C., March 17, 2016.

[2] U.S. Department of State, 2016.

[3] U.S. Department of State, *Ensuring Leadership in a Dynamic World: Quadrennial Diplomacy and Development Review*, Washington, D.C., 2015, pp. 57–58.

resources to organizations throughout the department. The department established CSCD in 2014 with QDDR-stated responsibilities that include "disseminat[ing] case studies, includ[ing] them in training exercises, and integrat[ing] them into interagency and communitywide planning."[4] Drawing on diplomatic reporting, open-source materials, interviews with practitioners, and roundtable discussions, CSCD examines different approaches to diplomacy to capture lessons learned and best practices on issues of relevance to the department. CSCD's analyses intend to help prepare American diplomats for the challenges they will face as they advance American interests around the world. The lessons identified are currently used primarily to improve instruction materials and approaches for FSI training courses. The work of CSCD and other ongoing efforts in the Department of State is a starting point from which to build an expanded lessons-learned program.

Background on Lessons-Learned Theory

> [T]he more effective organizations are at learning the more likely they will be at being innovative or knowing the limits of their innovation.[5]

Lessons-learned practices can facilitate an organization's ability to innovate and adapt by identifying the changes needed to achieve or sustain desired organizational effects. At times, these desired effects require sustaining what works well and/or changing what does not work efficiently and effectively or that has too much risk. Nick Milton, an expert in lessons learned, asserts that "A lesson learned is a change in personal or organizational behaviour, as a result of learning from experience."[6]

Organizational culture has a major influence on how well an organization can adapt to change and become a learning organization. Edgar Schein, a prominent organizational development theorist, provides a definition for organizational or group culture that is tied to learning:

> The culture of a group can be defined as a shared pattern of assumptions learned by a group as it solves its problems of external adaptation and internal integration, which has worked well enough to be considered valid and, therefore, to be taught to the new members as the correct way to perceive, think and feel in relation to those problems.[7]

---

[4] U.S. Department of State, 2015, p. 58.

[5] Chris Argyris, *On Organizational Learning*, Oxford, U.K.: Blackwell Business, 1999, p. xv.

[6] Nick Milton, *Lessons Learned Handbook: Practical Approaches to Learning from Experience*, Oxford, UK: Chandos Publishing, 2010, p. 16.

[7] Edgar H. Schein, *Organizational Culture and Leadership*, 4th ed., San Francisco, Calif.: Jossey-Bass, 2010, p. 22.

Schein's perspective on culture is relevant to the present discussion for several reasons. First, his definition of culture focuses on patterning and integration. Groups accumulate learning through two primary mechanisms: (1) size, growth, and adaptation to factors in the external environment and (2) internal integration that permits daily functioning and the ability to adapt and learn. Schein further contends that culture forms as groups constantly strive to pattern and integrate based on *a history of experiences or learnable events*. Thus, one of the major challenges often encountered in efforts to shift organizational culture is the lack of shared or accumulated learning.

This report provides insights for how the Department of State can set the optimal conditions for a culture of learning as it expands its institutional lessons-learned efforts. We offer insights on the crucial role that lessons-learned efforts will play in changing departmental culture to meet the needs of next-generation diplomacy. Ultimately, the aim is to increase the culture of learning and sharing of experiences within an organization.

Although Milton is one of the few experts who addresses how to employ lessons-learned approaches from a research-based or academic perspective, there is considerable literature on the application of lessons learned in specific practitioner communities. Consequently, practitioner "how-to" products also exist, and we cite some of them in this report. We have also sought to integrate the two perspectives, academic and practitioner, in identifying optimal approaches to employing lessons-learned activities. In addition, although we approached lessons learned from the perspectives of organization management and learning, we drew from a number of theoretical communities that intersect with these fields, including organizational and corporate learning, knowledge management, adult education, philosophy, failure theory, and management.

Chris Argyris, late business theorist and professor emeritus of Harvard Business School, provided seminal input to the merged approach of looking at lessons learned through the lenses of organizational theory and learning through his pioneering work in the 1990s on organizational learning. Argyris centered his work on the notion that organizational learning is a competence that all organizations should develop. He contended that the better organizations are at learning, the more likely it is that they will be able to detect and correct errors and to see when they are unable to detect errors.[8] He also believed that the more effective organizations are at learning, the more innovative they are likely to be. Argyris' latter premise is directly applicable to the Department of State's goals to be more innovative.

On the issue of error, bureaucracy exists in part to mitigate and, ideally, eliminate failure or error, but human error is a fact of life that must be acknowledged before a lessons-learned program can thrive. Argyris defines an error as "any mismatch between plan or intention and what actually happened" and notes that organizations have the most difficulty learning when problems are difficult, embarrassing, or threat-

---

[8]   Argyris, 1999, pp. xiii–xv.

ening.[9] Consequently, it is from exactly these types of errors that an organization's most important lessons can be learned. British Petroleum's learning and improvements following the Deepwater Horizon disaster of 2010 will affect not only the company's own personnel safety, investor confidence, and environmental impact but also oil rig safety across the industry. The failure of the United Parcel Service (UPS) to deliver packages on Christmas Day in 2013 was a public embarrassment, but the company's extensive lessons-learned investigation allowed it to make adjustments to improve services and maintain customer trust. The magnitude of error in these examples is very different, but there is always something to be learned from such large and even from seemingly inconsequential errors.

Kathryn Schulz—author, journalist, and staff writer at *The New Yorker*—has argued that "wrongness is a vital part of how we learn and change" and that, ideally, errors lead to a better understanding of oneself and of the world in general.[10] Applying this concept to organizations, such as the Department of State, shows that it *is important for organizations to recognize that errors will occur* and to prepare for them by establishing and maintaining a lessons-learned capability to help the organization improve. "Given the centrality to our intellectual and emotional development [think of this in terms of the department's need to evolve in response to a constantly changing world], error shouldn't be an embarrassment, and cannot be an aberration."[11] These ideas suggest that a truly effective organization has the internal processes necessary to face error with resiliency and emerge from such processes improved.

Purpose and Research Questions

To bolster the Department of State's efforts to expand its lessons-learned activities, we identified best practices for lessons-learned initiatives across government, the private sector, and academia. Toward that end, this research provides a theoretical backdrop and drawing on the field of organizational theory, an exploration of best practices for developing a strong culture of learning and implementing lessons-learned programs. We do not assess or otherwise comprehensively measure the department's current lessons-learned capabilities and do not prescribe a tactical implementation plan. Instead, we focus on best practices that can inform the department's next steps in furthering its enterprise lessons-learned initiatives.

The following questions guided our research and highlight the Department of State's key interests regarding best practices in lessons learned:

---

[9] Argyris, 1999, p. xiii.

[10] Kathryn Schulz, *Being Wrong: Adventures in the Margin of Error*, New York: Ecco, 2010, p. 5.

[11] Schulz, 2010, p. 5.

- How do organizations undertake lessons-learned efforts? How do they ensure that such efforts are comprehensive, fact-based, and free from undue bias or influence? What research and evaluation techniques are effective and ineffective? What organizational structures make such efforts most effective?
- How are lessons-learned efforts managed and overseen in such a way that employees see them as organizational priorities and as performance-enhancing exercises, instead of as efforts to assign blame for mistakes?
- How do organizations use the results of lessons-learned studies to change practices and behavior? How are lessons incorporated into professional training?
- How, if at all, do efforts to study and apply lessons learned differ in the public and private sectors?

Methods

To address these questions, we conducted a literature review, interviews, and round-tables. Appendix B discusses how we integrated the information from these research streams to identify best practices. To ensure that the best practices identified were relevant to the Department of State, we also familiarized ourselves with the department's organization and existing institutional and departmental lessons-learned activities.

Literature Review

We reviewed the extant theoretical literature on lessons learned and related concepts across various communities. For example, we examined academic sources related to lessons learned, corporate learning, organizational theory, and related subject areas. We also gained access to the Department of Defense (DoD) Joint Lessons Learned Information System (JLLIS) to explore archived postings of lessons-learned reports shared by groups within DoD to see what types of information were accessible, how accessible that information was, and what types of users were accessing the system.

We also reviewed the literature on key topics that directly influence an organization's ability to establish a lessons-learned program. Working from expertise within our team and a preliminary investigation of lessons learned, we explored such topics as lessons-learned theory and practice, organizational learning and corporate learning, business process management (BPM), and knowledge management. Having determined the fields that directly influence establishing a lessons-learned program, we deliberately anchored the findings of the literature review and interviews to academic theory. We provide insights from leading theorists and experts from the aforementioned fields to provide the foundation for the areas we emphasize throughout the report.

Interviews

Either the research team and/or FSI leadership contacted organizations for participation in structured or unstructured interviews. The selected organizations varied in size, type, and mission to allow our findings to be broadly applicable. Following any required approvals and/or executive prescreens, we conducted interviews through site visits or teleconferences with representatives of the following organizations (those with asterisks asked not to be identified by name):

- Private industry
  - a global information technology (IT) company*
  - Bloomingdale's
  - a global insurance company*
  - a large health-care organization*
  - a Six Sigma consultant*
  - an oil company*
  - a global hotel company*
  - Brigham and Women's Hospital
  - a multinational oil company*
  - an international aid nongovernmental organization*
  - Beth Israel Deaconess Medical Center
  - UPS
  - Bechtel
  - Psychological Associates
  - an international bank*
  - General Electric (GE)
- U.S. government (non-DoD)
  - the Federal Emergency Management Agency (FEMA)
  - the Government Accountability Office's (GAO's) Acquisition and Sourcing Management team
  - the U.S. Agency for International Development (USAID)
  - the Central Intelligence Agency's (CIA's) Center for the Study of Intelligence (CSI)
  - the National Aeronautics and Space Administration (NASA)
  - the following Department of State sections
    - CSCD
    - FSI Curriculum
    - Diplomatic Security
    - Transparency Coordination
    - Operations Center, Office of Crisis Management and Strategy
- U.S. military
  - the North Atlantic Treaty Organization (NATO)

- U.S. Army, Combined Arms Support Command
- U.S. Marine Corps (USMC)
- U.S. Air Force (USAF)
- Navy Warfare Development Command
- U.S. Coast Guard
- the Center for Army Lessons Learned (CALL)
- Joint Staff J-7, Joint Force Development
- Joint and Coalition Operational Analysis Division, Joint Staff J-7, Joint Force Development
- the Center for Complex Operations (CCO), National Defense University
- the Joint Center for International Security Force Assistance
- U.S. special operations forces
- the Ebola Task Force.

Several factors drove our data collection in the field. We foremost wanted to interview individuals who are involved in any aspect of lessons learned from a variety of organizations to validate common aspects of such programs. This array included organizations from different industries, of differing sizes, and with differing governance structures. We also looked at organizations that manage different levels of operational risk, such as the medical community, NASA, oil companies, and the construction industry. Furthermore, with FSI, we selected organizations based on their level of proficiency in lessons learned. Bechtel has an award-winning lessons-learned program, as does GE. The U.S. military also provides exemplary lessons-learned practices. We also selected organizations that developed or enhanced their lessons-learned capabilities following public crises or triggers that led to negative attention. Some organizations had attributes or challenges similar to those of the Department of State, such as those related to global operations. We sought interviewees from organizations with high turnover or rotational assignment structures similar to that of the Department of State's Foreign Service, which rotates assignments regularly. We selected other organizations, such as the CIA, because of similar institutional training culture and practices.

Of the 45 organizations that we contacted, 35 agreed to participate. We culled both best and worst (or poor) practices from organizations, regardless of size, type, or mission. While we targeted some organizations because of the similarity of their experiences to those of the Department of State, we interviewed others that had little resemblance but much to offer. Some organizations asked that we not identify them by name, but all were forthright in offering their successes and failures in lessons-learned endeavors. We believe this speaks to the supportive nature of these organizations and their awareness and respect for the Department of State's important global mission. Appendix A contains the interview protocol, and Appendix B details how we manually coded the interview notes and organized key findings that emerged from interviews, broken out across military and public- and private-sector organizations.

One limitation of our approach is that we drew conclusions from small samples of organizations in the various fields to ensure broader applicability of the findings; however, these sample organizations may not necessarily be representative of their organizational forms. Larger samples should be considered to develop a more comprehensive understanding of the trends in lessons-learned initiatives inside a specific field.

Roundtables

The research team also conducted two roundtables at the FSI campus in Arlington, Virginia, in partnership with FSI and CSCD. The first roundtable, conducted on September 14, 2016, showcased best practices in lessons learned in the military through collaboration among the Department of State, the military, and the research team. The roundtable included participants from National Defense University, the USMC, the USAF, the U.S. Coast Guard, and the Joint Staff. The second roundtable took place on October 20, 2016, and focused on best practices in lessons learned in the private industry and nonmilitary government organizations. Participants included representatives from Conoco Phillips, UPS, the CIA, FEMA, and NASA.

Organization of This Report

The remainder of this report describes the results of our research, which we have organized around the construct of what, when, why, who, and how (see box). Chapter Two discusses *what* a lessons-learned program is and provides examples of *when* lessons-learned activities could or should happen. The concept of organizational learning

---

Questions Guiding the Organization of This Report

What?   Describe the essential steps in a lessons-learned process.

When? Examine how organizations know or decide when to undertake a lessons-learned review.

Why?   Provide frameworks to help organizations determine what the implementation of a lessons-learned program aims to achieve. We refer to this as *vision*.

Who?   Map the interrelationships of the individuals and organizational subcomponents involved in lessons learned.

How?   Outline options for strategic approaches to lessons learned and describe associated enabling infrastructures.

emerged in the 1990s through the work of Chris Argyris, and the Department of State's interest suggests a wider trend of growing curiosity.[12] A policymaker who is curious about a new organizational concept may signal an openness to apply such concepts to his or her own organization, and it is vital to ensure that the concept is well defined before it is discussed as a tool to achieve the policymaker's vision. To that end, Chapter Three addresses *why* organizations should employ lessons-learned activities. Chapter Four examines *who* is involved. Chapter Five then discusses *how* organizations are conducting their lessons-learned efforts, including associated key enablers, incentive programs, and knowledge management, among other issues essential to a strong lessons-learned program. Chapter Six discusses our findings and thematic best practices for developing lessons-learned programs applicable to the Department of State.

---

[12] The first edition of his work, *On Organizational Learning*, was published in 1992.

CHAPTER TWO

The *What* and *When* of Effective Lessons-Learned Programs

This chapter defines and outlines the goals and purposes of a lessons-learned program as undertaken in a variety of organizations. *What* are the essential steps in a lessons-learned process? *When* should organizations undertake a lessons-learned review? Chapter Three addresses potential outcomes of lessons-learned programs with respect to the vision of the organization.

The literature and the practitioner community broadly concur that a lesson is not *learned* unless or until something changes as a result. While there are many definitions for *lessons learned*, all are tied to the notion of changed behavior as a result of learning from experience.

Some academic practitioners describe a *lesson learned* as Milton puts it:

> a change in personal or organizational behaviour, as a result of learning from experience.[1]

CALL defines a lesson learned as

> validated knowledge derived from actual experience, observation, and analysis of military training and actual operations that results in changed behavior by Soldiers, leaders, and units.[2]

NATO's Joint Doctrine for Operations describes the purpose and requirements of a lessons-learned program as follows:

> to learn efficiently from experience and to provide validated justifications for amending the existing way of doing things, in order to improve performance, both during the course of an operation and for subsequent operations. This requires lessons to be meaningful and for them to be brought to the attention of the appropriate authority able and responsible for dealing with them. It also

---

[1]   Milton, 2010, p. 16.

[2]   CALL, *CALL Services*, Fort Leavenworth, Kan., Handbook 15-11, June 2015, p. 5.

requires the chain of command to have a clear understanding of how to prioritize lessons and how to staff them.[3]

We built on both the theory and practitioner approaches to produce a graph-
ical depiction of the lessons-learned cycle at its most basic level, which consists of
three phases (Figure 2.1).[4] During the *action and analysis* phase, a "learnable event" is
identified and analyzed. This analysis produces *lessons identified*, which are validated
by key stakeholders (or recycled to the analysis stage, if necessary). Once validated,
practitioners disseminate and institutionalize lessons identified during the *implement*
stage. Implementation both completes and restarts the lessons-learned cycle; a lesson
is not "learned" until a change is implemented. Continuous monitoring and evaluation underpin the lessons-learned cycle—because practitioners may identify additional
learnable events at any point in this process.

The vast majority of organizations we interviewed, whether in the public, private,
or military sectors, follow the general process outlined in Figure 2.1, albeit with organizational nuances and adjustments. The remainder of this chapter discusses the three
phases in greater detail.

Figure 2.1
Lessons Learned: A Continuous Process

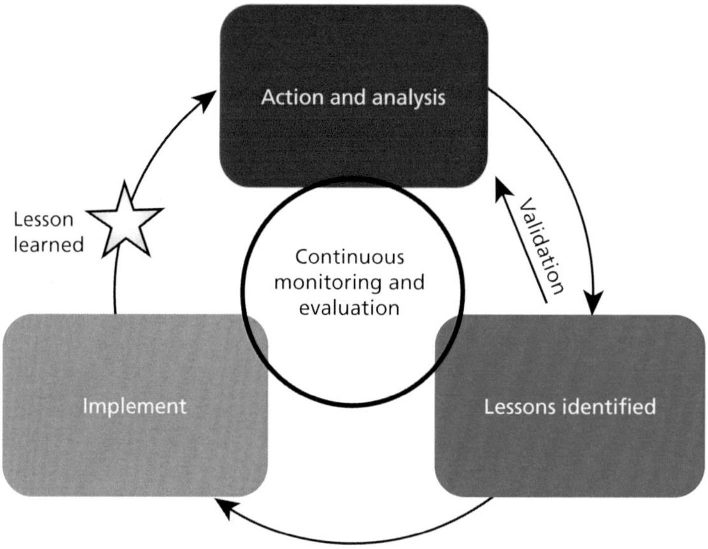

RAND *RR1930-2.1*

---

[3] NATO, *Allied Joint Doctrine for the Conduct of Operations*, AJP-3(B), Brussels, March 2011, Para. 0454.

[4] Milton defines three broad steps in what he calls *the learning loop*: activity, lessons identified, and updated
documents and processes (Milton, 2010, p. 16).

Action and Analysis

The action-and-analysis stage of the lessons-learned process entails the identification of learnable events, data capture, analysis of data, and validation of initial findings.

Identifying Learnable Events

Learnable events are events, conditions, or outcomes that trigger a lessons-learned review. An organization's vision and mission should serve as the primary determinants of what constitutes a learnable event; these priorities help lessons-learned practitioners scope and categorize topics, events, processes, or triggers.

Strategic Collection Planning

Strategic collection planning gives lessons-learned practitioners a framework for identifying learnable events and determining when to conduct lessons-learned reviews. However, it also helps focus the organization on key issues (and prioritize what needs to be developed as a lesson learned) and can foster senior leader and stakeholder buy-in during the early stages of a lessons-learned review. Organizations adopt a range of parameters for determining triggering events or conditions. While some organizations determine a set of conditions or characteristics that should trigger a lessons-learned review (e.g., incidents of failure or success), other organizations collect data continuously for lessons-learned processes as part of their normal business practices—or even develop an annual collection plan.

No matter the procedure for or approach to determining what events are learnable, the organization's mission and vision must inform and drive what triggers a lessons-learned review. The enterprise's vision will track throughout the entire lessons-learned process: The mission statement helps define goals and objectives, which informs the strategy, which leads to the implementation plan, which has metrics for measuring progress that tie back to the organizational mission and vision.

Reactive Approaches

One means of identifying learnable events is to determine a set of characteristics for an incident to trigger a lessons-learned review. These triggers need not only be points of failure, such as cost overruns or safety and security violations, but can also be successes, such as successful diplomatic negotiations or the streamlining of business processes at an embassy. Milton explains that these triggers may be determined in advance or can be flagged by lessons-learned experts seeking "opportunities for lesson identification."[5] The U.S. Coast Guard, for example has a lessons-learned model that responds to events. It has created an incident management handbook that "assists Coast Guard personnel in the use of the National Incident Management System's Incident Command

---

[5]   Milton, 2010, p. 33.

System during response operations and planned events."[6] The U.S. Fire Administration suggests "predesignating specific types of incidents that would automatically trigger a [formal] critique," for example, fires with injuries, fatalities, and/or exceeding a predetermined dollar amount of damage.[7] A similar approach for the Department of State might pertain to safety and security incidents at embassies.

One of the advantages of this approach to identifying learnable events is its flexibility and ability to react to unforeseen—but important—events. For example, remote or deployed employees may be exposed to a potential learnable event that the central office is unaware of or that the central office cannot react to in a timely manner. A weakness, however, is that collection may be post hoc and inconsistent, or even duplicative.

Scheduled Approach

A second approach to lessons-learned reviews entails scheduled collection. This approach seeks to inculcate lessons-learned processes into an organization's normal business processes or, as one lessons-learned practitioner described, "to put the fluoride [lessons-learned processes] into the drinking water."[8] This approach makes lessons-learned reviews integral to the way the organization does its work. For example, project managers for a large hotel chain we interviewed outbrief at the close of new hotel developments, explaining whether targets were achieved (in multiple stages); lessons are captured and fed into a project management database for future project managers.[9]

Organizations may also adopt a hybrid approach that incorporates the two approaches. For example, FEMA plans a lessons-learned agenda in advance to match agency priorities and strategies to the greatest extent possible, but also tackles incidents on an ad hoc basis (see box on next page for an example). The U.S. Navy routinely conducts reviews at three stages: predeployment at the end of training, midcruise, and at the end of the cruise.[10]

A number of organizations we interviewed develop annual lessons-learned collection plans under the guidance of senior leadership, including the vast majority of military-sector organizations. These annual plans can look both forward and back, addressing future goals and priorities and recently discovered weaknesses. Organizations seek to identify and validate the existence of knowledge gaps in critical areas,

---

[6]   Commandant Publication P3120.17B, *U.S. Coast Guard Incident Management Handbook*, May 2014.

[7]   U.S. Fire Administration, "Special Report: The After-Action Critique: Training Through Lessons Learned," USFA-TR-159, April 2008, p. 3.

[8]   Comments at the RAND-CSCD Lessons Learned Roundtable held at the FSI campus, Arlington, Va., September 14, 2016.

[9]   Interview with the chief learning officer (CLO) of a hospitality company, McLean, Va., May 5, 2016.

[10]   Telephone interview with Director, Navy Lessons Learned, June 23, 2016.

Learnable Events at Any Stage

As Hurricane Matthew approached the United States in late 2016, FEMA initiated collection efforts to capture learnable events over the course of the agency's hurricane response. As it was collecting learnable events as part of its established lessons-learned program, the agency also began examining its lessons-learned processes in real time—capturing learnable events not only on hurricane response but also on lessons-learned business processes. In an interview, agency representatives referred to this as "lessons learned on lessons learned," showing that, indeed, there are learnable events everywhere there is activity.

---

focusing on behaviors or tactics that were successful or problematic rather than on people who were successful or problematic.[11]

Annual collection planning has several key benefits. First, the involvement of and buy-in from senior leadership lend credibility to the resultant lessons-learned efforts. Senior leader involvement can provide "top cover" to collection teams that need to gain access to people, documents, and information to capture data. Second, senior leader buy-in can help drive financial and human resources to lessons-learned efforts. During interviews,[12] all military representatives mentioned the importance of senior leader buy-in and involvement; USMC and USAF representatives, in particular, stressed the essential nature of senior leader buy-in and involvement at *all* stages of lessons-learned reviews.

Annual collection plans may develop a set list of studies or issue areas to address the following year. Narrowly scoping annual collection plans can help mitigate resource constraints. The Marine Corps Center for Lessons Learned—which, like the CSCD, has finite resources—develops an annual collection plan to identify a limited number of topics worthy of study.[13] Whether conducted on an annual or rolling basis, continuous, scheduled collection integrates lessons-learned processes into organizational processes, practices, and culture. When conducted on a regular basis, lessons-learned reviews are no longer standalone processes but become part of core business practices at multiple stages of projects and become integral to the way an organization does its work.

---

[11] Mark White and Alison Cohan, "A Guide to Capturing Lessons Learned," Nature Conservancy, undated, p. 11.

[12] See Appendix B for more detail.

[13] Interview with Head of Collection and Analysis Section, Marine Corps Center for Lessons Learned, Quantico, Va., May 23, 2016.

Data Capture

Lessons-learned reviews use a variety of methods and approaches to capture data for analysis. Tools for data collection include questionnaires, checklists, interviews, reviews of documentation, observation, focus groups, workshops, team debriefs, and after-action reviews (AARs). Different data-collection methods and sources will require different types of people to conduct the review and capture the data—an area we delve into in Chapter Four.

In this subsection, we will explore three examples of ethnographic research that provide unique contributions to lessons-learned processes: interviews, observations, and AARs.[14] We will also outline the advantages and disadvantages of these approaches to capturing lessons-learned data. For more detail on other methods of data collection, Figure 2.2, produced by The Partnering Initiative, describes and addresses the advantages and disadvantages of a range of additional data-collection methods, including questionnaires, surveys, and checklists; documentation reviews; focus groups and workshops; and reviews beyond the documentation, allowing the group to assess its performance and the decisions it has made.

Interviews—structured, unstructured, or semistructured—are often at the core of a lessons-learned review data-capture process. White and Cohen offer four principal questions an interview should address: (1) What went well? (2) What did not go well or had unintended consequences? (3) If you had it all to do over again, what would you do differently? and (4) What recommendations would you make to others doing similar projects?[15] Active listening—including suspending judgment, avoiding interrupting, paraphrasing to check for understanding, and stating the purpose of questions—helps ensure an open conversation and improves the quality of interview outputs.[16]

Milton describes interviews as "the most effective way to identify lessons from a single person,"[17] and using interviews to capture data for lessons-learned processes has a number of advantages. Most important, interviews can provide a critical diversity of perspectives; people in different roles will have different experiences, concerns, takeaways, and insights.[18] However, lessons-learned practitioners should also recognize certain risks of or drawbacks to relying on interviews alone. While interviews can add nuance and a diversity of perspectives, they can also be time consuming and expen-

---

[14] The University of Cambridge defines *ethnographic research* as research that "aims to produce a detailed description of how a particular social group operates, based on observation of, and often participation in, the group. This may be supplemented by interviews and gathering of documents and artefacts" (University of Cambridge, "Ethnographic and Field Study Techniques," web page, undated).

[15] White and Cohen, undated, pp. 6–7.

[16] For more detail, see Milton, 2010, pp. 59–61.

[17] Milton, 2010, p. 59.

[18] Telephone interview with senior director of Strategy and Learning, Mercy Corps, June 1, 2016.

Figure 2.2
Data Collection: Methods and Sources

TOOL 5
Data Collection: Methods and Sources

METHOD	PURPOSE	ADVANTAGES	DISADVANTAGES
Questionnaires, Surveys and Checklists	■ Usually devised and administered to obtain statistical data for a particular question or set of questions. ■ Often associated with quantitative research. ■ Used when you need to get information quickly and/or easily from people in a non-threatening way.	■ Can be completed anonymously ■ Inexpensive to administer ■ Easy to compare and analyse responses ■ Can be sent to a large number of people	■ Might not get a detailed response ■ The wording can cause bias ■ Impersonal nature ■ May not include all information ■ Might have a low response rate
Structured, Semi-structured and unstructured interviews	■ Useful to obtain a fuller understanding of someone's impressions or experiences of the partnership or to delve into more details about questionnaire responses. ■ Structured interviews tend to be used for quantitative research; semi-structured and unstructured interviews tend to be used in qualitative research.	■ Collects a full range and depth of information ■ Can be flexible with interviewees	■ Time consuming ■ Can be hard to analyse and compare ■ Interviewer can bias responses ■ Can be costly if involves face-to-face interviews
Review of Documentation	■ Conveys information about how the partnership operates. ■ Documents can include: MOUs, web literature, meeting minutes, films, partnership agreements, etc.	■ Collects comprehensive and historical information ■ Is not disruptive to the partnership ■ Information already exists ■ Less rooms for biases in interpreting the information	■ Can take time ■ Information can be incomplete or out of date ■ Need to be clear about what you are looking for ■ Not a flexible approach
Participant and Direct Observation	■ To gather accurate information about how a partnership operates. ■ Participant observation requires the partnership researcher to become a participant in the culture or context being observed. ■ Direct observation involves the researcher observing actual situations or interactions rather than being told about them.	■ View operations of a partnership as they are occurring ■ Can adapt the case study in accordance with the events as they happen	■ Can be difficult to interpret observed behaviours ■ Can be complex to categorise observations ■ Observer can influence behaviour of partnership participants ■ Difficult to remain impartial if participating ■ Can be expensive
Focus Groups / Workshops	■ Involve organised discussion with a selected group of individuals to gain information about their views and experiences about a topic. ■ Usually involves exploring a range of views or a topic in depth through discussion.	■ Reliable sources of impressions that are shared by all ■ Can be an efficient way to get a broad range and depth of information in a short time ■ Can convey key information about the partnership	■ Can be difficult to analyse responses ■ Needs a good facilitator for safety and closure ■ Can be difficult to schedule people together
Reviews	■ Provides an opportunity for partners to reflect on the value of the partnership, determine whether the partnership is meeting its desired objectives. ■ Offers a chance to agree as a group to any revisions to the partnership agreement.	■ Allows opportunity to collect information from all partners ■ Can allow for a deep analysis of the partnership ■ If skilfully done can be a significant catalyst for improving the partnering process and relations.	■ Can be time consuming (especially if reviewer is meeting with different partners) ■ Most effective after the partnership has been operating for some time ■ An external reviewer could potentially be destructive to the partnership

SOURCE: The Partnering Initiative, *The Case Study Toolbook*, August 2014, p. 71. Used with permission.
RAND *RR1930-2.2*

sive, difficult to compare, and risk inserting interviewer biases. [19] Additionally, while a number of lessons-learned processes rely on unstructured or semistructured interviews, this approach may not work for all types of reviews. The U.S. Fire Administration cautions: "Proceed systematically" and do not "permit unstructured, meandering, disorganized discussions."[20]

Lessons-learned reviews may also capture data via direct or participant observation of learnable events—as opposed to postevent data collection. This form of ethnographic field research often involves firsthand data collection as augmented by interviews of other participants. CALL employs this form of observation, deploying representatives with active units. FEMA also deploys lessons-learned officers alongside incident response teams to undertake real-time observation and collection.[21] As with interviews, observation-based ethnographic research can add nuance and a diversity of perspectives. However, it can also be expensive and time consuming and risks inserting observer biases and influencing the behavior of participants being observed.[22] Information or data overload can be a challenge. Many interviewed organizations, such as FEMA and Mercy Corps, sought to limit the number of observations. Referring to observations, Avis and Sharpe suggest that "[f]ive is much better than 500."[23]

Finally, participants in a project or activity can use an AAR to identify lessons learned. AARs are common in government organizations, particularly the military. According to Milton, AARs "are short, structured, review meetings, conducted to draw out lessons from a task or activity."[24] AARs usually are recorded in the form of a brief report. The box on the next page briefly describes FEMA's AAR format.

An important strength of AARs is that they allow those directly involved in an event the opportunity to reflect and identify both successes and failures in "an open, blame-free, inclusive environment" with established ground rules to mitigate any reluctance to share.[25] One weakness of AARs is that they often contain information that might identify an individual at fault, and this can lead to feelings of vulnerability and reluctance to use this form of review. While AARs are widely used in the U.S. Army, these often have little public visibility; for example, an AAR examining a training activity would not be public. The Department of State operates in a public and

---

[19] The Partnering Initiative, 2014, p. 71.

[20] U.S. Fire Administration, "Operational Lessons Learned in Disaster Response," Emmitsburg, Md.: Federal Emergency Management Agency, June 2015, p. 51.

[21] Interview with FEMA Assistant Administrator, Washington, D.C., June 13, 2016.

[22] The Partnering Initiative, 2014, p. 71.

[23] Peter Avis and Joe Sharpe, "Operationalization of the Lessons Learned Process: A Practical Approach," in Susan McIntyre, Kimiz Dalkir, and Irene C. Kitimbo, eds., *Utilizing Evidence-Based Lessons Learned for Enhanced Organizational Innovation and Change*, Washington, D.C.: IGI Global, 2015, p. 78.

[24] Milton, 2010, p. 54.

[25] Milton, 2010, p. 58.

FEMA's Robust Use of AARs

FEMA responds to emergencies. Learning opportunities are vast in this type of work, and FEMA's use of AARs is robust by necessity. We examined several FEMA AARs and found them to brief and to follow a standard format. They provide a description of the event, a fact sheet, a discussion of unresolved critical issues of national importance that would affect future operations, reference maps and charts, and organizational charts of the response team.

---

often sensitive political environment. Many department employees may be reluctant or unwilling to participate without proper protections.

Analysis and Validation

As soon as data collection begins, lessons-learned teams should begin analyzing and validating initial findings to hone and refine data collection. While interviewees had little to say on "how to *analyze* data," they provided a wealth of information on the importance of validating data and initial findings and on how to approach and undertake validation in lessons-learned processes.

Interviewees stressed the importance of validating findings to ensure that the information gathered is comprehensive and that the analysis is as unbiased as possible. Lessons identified need to be validated to ensure they are still relevant and worthy of being archived and/or disseminated.[26] The literature confirms the essential nature of a simple but effective validation process.[27] White and Cohen explain the need to verify the accuracy and applicability of lessons identified and assess whether they are locally or broadly applicable.[28] Milton reminds practitioners of "the need to remove the 'garbage in, garbage out' problem that besets so may lessons learned systems," through validation and quality control.[29]

Validation techniques include triangulation across multiple sources, review, and validation by subject-matter experts (SMEs) and both centralized and decentralized vetting processes. Resource constraints can limit validation capacity and approaches. In part because it is relatively small and has resource constraints (a challenge CSCD shares), the Marine Corps Center for Lessons Learned centralizes both analysis and

---

[26] Telephone interview with lessons learned coordinator at oil company, May 4, 2016.

[27] Milton, 2010; Mark Marlin, "Implementing an Effective Lessons Learned Process in a Global Project Environment," *Annual Project Management Symposium Proceedings*, Dallas, Tex., 2008; and Art Murray and Jeff Lesher, "The Future of the Future: Breaking the Lessons-Learned Barrier," *KM World*, August 31, 2008.

[28] White and Cohan, undated, p. 2.

[29] Milton, 2010, p. 71.

vetting. Other organizations are able to stand up lessons-learned committees or facilitators that review, validate, and assess lessons learned, including identifying differences between what was promised and what is being delivered.[30] Many organizations take the middle road, with less labor- or resource-intensive validation processes. For example, at Brigham and Women's Hospital, after interviewing all concerned and writing up findings, the Patient Safety Team validates the report with all involved.[31]

Although—and in part because—validation is a crucial step, the process must be straightforward and as simple as possible. Murray and Lesher advocate simplifying and streamlining the validation process "commensurate with risk," or "people quickly become frustrated with the formal system and return to the simple water-cooler methods of knowledge sharing."[32]

Lessons Identified and Documented

Validated findings from the action-and-analysis phase become lessons identified: "A recommendation, based on analyzed experience (positive or negative), from which others can learn in order to improve their performance on specific task or objective."[33]

Lessons identified should be documented in ways tailored to their target audiences, presented in a way that is easy to follow and well-structured and that provides an appropriate but concise amount of context to achieve maximum impact.[34] Individuals who administer successful lessons-learned programs take into account the most effective and efficient methods of delivering the findings to the appropriate audience. Senior leadership may respond better to a short formal briefing, white paper, or slide. Implementers may benefit most from curricula or training; policy analysts may benefit most from a report.

Deliverable formats include long-form reports, briefings, and presentations; white papers; AARs; inter- and intraagency databases; and new curricula, training, and doctrine. Regardless of the format of the deliverable, it should provide a clear, concise summary, as well as actionable and measureable recommendations that stakeholders can implement. How the material is presented also matters, and storytelling is often an effective approach. "[C]ompelling stories help people understand," which makes it easier to acknowledge shortcomings and identify solutions.[35] Reports should be well

---

[30] Telephone interview with lessons learned coordinator at oil company, May 4, 2016.

[31] Telephone interview with health care officials, May 18, 2016.

[32] Murray and Lesher, 2008.

[33] Milton, 2010, p. 36.

[34] Milton, 2010, pp. 67–72.

[35] Telephone interview with healthcare executives, May 18, 2016.

written and easy to digest. Having a single drafter may help ensure unity of voice. [36] Finally, ease of use is important: End users should receive training and instruction on how to take advantage of database resources, and databases should provide metadata for end-user ease of use and searchability.[37] For example, FEMA compiles operational "observations" into a database and tags or indexes each observation with metadata to allow easy sorting and cataloging.[38]

Successful organizations also acknowledge and account for resource limitations, recognizing that the broader the audience, the more financial resources you should invest in products and that, likewise, products for smaller audiences should cost less. Accordingly, organizations with limited financial resources should consider narrowing their target audience. These differences and trade-offs are visible in the U.S. military's lessons-learned community of practice. Well-resourced CALL provides a full suite of products, including handbooks and personalized outputs, to a variety of stakeholders and potential implementers. The Marine Corps Center for Lessons Learned, on the other hand, still produces a sleek, robust product but targets only senior leadership, in acceptance of resourcing trade-offs. These organizations know their resourcing and target audience and have tied their strategy to their organizational vision and mission.

Implementation

Once organizations identify lessons, the next step is to institutionalize them. For organizations to learn lessons, not merely identify them, dissemination and implementation strategies should go beyond simply getting the right information to the right people. A NASA presentation emphasizes that practitioners should seek to "infuse lessons into procedures and training such that the project need not depend on the right person reading a lesson at the appropriate project milestone."[39] This stage is the most difficult; many interviewed organizations acknowledged this challenge and/or even identified a gap between lessons identified and lessons actually learned. Practitioners review, plan, and then act on the lesson identified. Among other examples, proposed actions might include changing doctrine, training, processes, or procedures and/or adjusting staffing. In accordance with the lessons-learned framework, attempted changes must be marketed, monitored, and then evaluated to feed back into the continuous cycle.

---

[36] David Oberhettinger, "Lessons Learned from Soup to Nuts," briefing, Pasadena, Calif.: Office of the Chief Engineer, Jet Propulsion Laboratory, California Institute of Technology, August 23, 2011, slide 5.

[37] Milton, 2010, pp. 108–109, outlines the types of metadata database entries should document, including lesson title; topic and subtopic; originator; and the date of the event; date of identification of the lessons; and/or date of documentation, action, lesson security classification, lesson status, and lesson value.

[38] Interview with FEMA assistant administrator, June 13, 2016.

[39] Oberhettinger, 2011, slide 7.

Dissemination

In general, dissemination efforts fall in one of two categories: passive, pull-type platforms and active, push-type activities. Platforms for passive dissemination rely on end users' seeking out information, and impact can be limited without marketing campaigns and other efforts to build awareness that the information is available. Websites, publishing reports, and knowledge management databases are all tools of passive dissemination. A number of organizations use passive platforms to highlight or disseminate their lessons-learned products. For example, CALL publishes many reports on its external website. Brigham and Women's Hospital publishes medical errors online in an effort to promote transparency.[40] Databases, such as the U.S. military's JLLIS, can serve as passive repositories for lessons. Agencywide or even interagency lessons-learned databases can empower individuals to seek out information without having to search multiple websites, repositories, or publications.

With active dissemination, an organization pushes lessons learned to the stakeholders who stand to benefit most and/or are best positioned to serve as key agents of change. Outreach to these change agents can take the form of targeted emails to individuals, announcements in internal publications and newsletters, databases that push relevant lessons-learned studies to select users, and passing along information or briefing books to successor staff.

Mercy Corps pushes lessons and information to staff via an online newsletter accessed by 96 percent of headquarters staff and 83 percent of field staff.[41] Similarly, one equipment manufacturer we interviewed would publish Six Sigma success stories in internal newsletters,[42] while an oil company reports lessons-learned in newsletters and via internal email.[43] The equipment manufacturer also has a centralized office that assesses whether a problem and solution an operational unit reports represents an enterprise-level challenge that should be disseminated broadly and result in broader process changes. If yes, the office disseminates lessons to other units though meetings, newsletters, databases, and training.[44] Finally, not all active dissemination is technology dependent: Representatives of a large hotel chain we interviewed described how managers disseminated lessons at daily shift kickoff meetings.[45]

A hybrid model combining both passive and active dissemination enables organizations to take advantage of the benefits of both approaches. Databases, in particular,

---

[40] Heather Punke, "Why Brigham and Women's Hospital Put Medical Errors in Blog Form," *Infection Control & Clinical Quality*, February 22, 2016.

[41] Mercy Corps, "Practicing What We Preach: A Review of Learning and Research Utilization," brief, Portland, Oreg., May 2015.

[42] Telephone interview with Six Sigma expert, April 27, 2016.

[43] Telephone interview with oil company lessons-learned expert, May 4, 2016.

[44] Telephone interview with Six Sigma, April 27, 2016.

[45] Interview with CLO of hospitality company, McLean, Va., May 5, 2016.

can provide both active and passive functions, allowing users to seek out information but also pushing content to users. For example, NASA's lessons-learned database has a subscription feature that emails lessons learned to staff.[46] One step further would be to individualize the content being disseminated based on user profiles, activities, and viewing patterns in a database. Chapter Five addresses IT-based databases in greater detail.

Geographic or functional differences may also influence dissemination of lessons. Organizations may tailor dissemination approaches to target different audiences, stakeholders, or environments. For example, when determining how to apply standard operating procedures in new or varying locations, FEMA must take into account different states' and localities' laws, administrations, and unique operating environments. A multinational oil company we interviewed employs multilevel information-sharing, in which functional managers identify successes that can be replicated in other regions of the world.[47] Company SMEs capture and upload lessons to SharePoint discussion threads as part of localized problem-solving. A centralized knowledge management staff curates and identifies lessons from these discussions, developing narratives for lessons with enterprisewide interest—i.e., those that are timely, globally relevant, and critical to business lines—and then posts them as articles on the organization's global portal. In a similar approach, managers at one hotel company roll up and disseminate the most compelling staff-generated guest-related lessons.[48] These approaches would be analogous to CSCD's processing and sharing transferable and scalable lessons identified at U.S. embassies.

Monitoring and Continuous Evaluation

Monitoring the implementation of lessons can provide helpful feedback on the success of lessons-learned processes, motivate adjustments to dissemination strategies, and serve to validate long-term lessons. Metrics for determining the impact of lessons-learned reviews vary depending on what needs to be measured (e.g., the quality of the deliverable or product, the rate of lesson production, staff or stakeholder engagement, safety, financial savings) but should be tied to the original organizational strategy and implementation plan. For example, FEMA has a coding system to evaluate the success of a new lesson learned. The "platinum" label indicates that a lesson allows the agency to save more lives in emergency response; the "gold" indicates that a lesson saves resources.

Even with established evaluation metrics, it can be very difficult to quantify and measure certain outcomes. When you are learning from social engagements, how can you determine whether something went well? In many cases, the process could be cor-

---

[46] Oberhettinger, 2011.

[47] Interview with Knowledge Advisor, multinational oil company, May 24, 2016.

[48] Interview with CLO of hospitality company, McLean, Va., May 5, 2016.

rect but the outcome undesirable. This is particularly true in the case of the Department of State: How do you know that you have conducted diplomacy well? In this scenario, monitoring whether a process was completed is a much more effective way to measure the impact of a lesson learned.

International companies, such as Bechtel and Coca-Cola, engaging with other countries first tap into archived history with a certain client or government to determine what issues arose and what was resolved. Preengagement activities that plan for the interaction, preparing for known questions, and understanding where hard lines exist are vital to the process. Finally, it is key to debrief the event, assessing what went well and what could be improved from both the process and outcome perspectives.

Ultimately, many of the interviewed organizations were still in the process of developing metrics to measure the impact of lessons learned. Many successes were anecdotal, but even anecdotal evidence can be a good starting point for proving the credibility of a lessons-learned program.

When done successfully, dissemination, implementation, monitoring, and continuous evaluation are important drivers of continuous lessons-learned processes. Together, these steps help nurture an organization's learning culture by normalizing and even lauding shortcomings. A strong culture of learning supports a continuous learning loop—important for successful lessons-learned programs.[49] As Milton explains, "[i]f we make sure that actions are fed back into activity . . . , then the learning loop is closed, activity will improve, performance will increase, and we will travel down the learning curve."[50]

Managing Shortfalls in Lessons Learned

A learning culture is different from what many may consider typical U.S. government bureaucratic culture. Bureaucracy depends on precedent and is difficult to change—sometimes major change even requires an act of Congress. In a learning culture, failure is accepted because it exposes weaknesses to be improved on. A failure becomes a learnable event to help the organization improve as a whole. An improved state of existence is not to be confused with a final state of existence, however. The achievements of a lessons-learned exercise will become obsolete in time. An improved process will eventually break down as the world changes around it. This is why successful organizations are reflective and adaptive.

The adaptive culture of a learning organization suggests that the lessons-learned process is a constant loop. If the organization identifies a failure or other shortfall, the organization determines how to fix it and implements a change that does so. Any subsequent failures of the change become lessons themselves. This is the reality of a

---

[49] Nearly every organization we interviewed noted the importance of cultivating a culture of learning. See Appendix B.

[50] Milton, 2010, p. 20.

reflective and adaptive organization—change is constant, and this constant change is sparked by some lesson for the organization to learn. A reflective and adaptive organization has accepted that "status quo" or an "achieved end state" will never exist again.

If, indeed, a lessons-learned exercise leads to a solution that makes matters worse, it might be caused by a shortfall in the lessons-learned process itself. The lessons-learned process is still a business process susceptible to shortfalls. A lessons-learned expert and knowledge management consultant has offered several "worst practices" in lessons learned:[51]

- Learn only from mistakes! (No, you can learn from successes and even by examining standard procedures.)
- Task one manager with the entire lessons-learned process—no need to involve the team! (No, the whole team is essential to identify lessons at all stages of a process.)
- Each lesson is unique, so there should be no standard format for a lessons-learned record or report! (No, standard formats make lessons easier to follow and remember.)

By avoiding such worst practices and following best practices, an organization will find that learning and improvement are, indeed, continuous. This means that improvements can be made at any level of an organization at any time—from daily processes at the most basic level to processes at the highest level where laws must be changed to turn a high-level lesson into practice. Broadly speaking, lessons-learned processes result in a change in mindset and a realization that status quo and precedent may have high risk for failure.

---

[51] Milton, 2010, pp. 175–184.

CHAPTER THREE

Why Implement a Lessons-Learned Program?

For an organization to avoid risk systematically and operate optimally, it must possess an internal system that allows critical thinking and flexibility to change.[1] Lessons-learned programs are one such tool for creating maximum efficiency and efficacy for countless corporations, military services, and government agencies. For such a program to be implemented well, the approach should be systematic and meet certain criteria at each step.

The first step toward creating a lessons-learned capability is for the organization to clarify the factors that drove stakeholders to explore it as an option for improved processes, then to determine aims of implementing a lessons-learned program.[2] Once these purposes are clear and translatable across the organization, the program is more likely to garner buy-in than if aims are ambiguous or seemingly impertinent to each unit or specific function.

Moreover, when purpose is linked to the broader organization, operational effectiveness is improved. If all levels of the organization are able to see that they can adopt lessons-learned processes to improve their output at both the unit and enterprise levels, individuals will be more able to see the value of a program that turns past mistakes and into future successes. An effective lessons-learned program is therefore both a means of operating more effectively at the enterprise or unit level, which complements the overall organizational vision.

Therefore, the success of a lessons-learned capability starts with a vision. Based on our literature review, data collected in the field, and expert opinion within our team, we determined that four key implementation decisions are essential in the vision of a lessons-learned program:

1. What are the desired outcomes of the program?

---

[1] Martha Grabowski and Karlene Roberts, "Risk Mitigation in Large-Scale Systems: Lessons from High Reliability Organizations," *California Management Review,* Vol. 39, No. 4, 1997.

[2] Centers for Disease Control and Prevention, "CDC Unified Process Practices Guide: Lessons Learned," Atlanta, Ga., November 30, 2006.

2. What approaches will be used to achieve these outcomes (e.g., BPM and/or corporate learning strategies)?
3. Whom do you wish to reach in the organization (individuals through leadership and/or business units)?
4. How will the lessons-learned program be governed (centralized or decentralized) based on desired outcomes and target audience?

In this chapter, we discuss the desired outcomes, target audience, and high-level organizational structure that need to be determined in a lessons-learned vision. We expand the discussion on organizational structure and target audience in Chapter Four and discuss the strategic approaches to lessons learned in Chapter Five.

Desired Outcomes

The anticipated goals and outcomes of the lessons-learned process must be clearly stated as part of the program vision from the outset. Desired outcomes will generally depend on the overall goals of a particular company or organization. For example, an organization that focuses on production may direct its lessons-learned program toward improving production quality and process efficiency, while an organization staffing employees in warehouses to work with heavy machinery may emphasize safety.

Laura Lane, the president of Global Affairs of UPS, shared how the organization improved its lessons-learned practices to achieve better outcomes. She reported that UPS realized it needed to change the focus of its lessons-learned program from measuring events to strategies that would improve results. As a result of this change in mindset, the current program focuses on measuring goals and outcomes of plans made at all levels of the organization well in advance; the metrics for success are based on quantified profit items set forth in the company's vision statement. While this practice may not work for the Department of State, it does illustrate that an organization should have a vision for its lessons-learned program that nests into the overall vision of the organization.

Although goals and processes among organizations vary considerably, we observed four broad reasons for organizations to implement lessons-learned programs:

- to become a learning organization
- to mitigate risk
- to increase efficiency
- to maintain institutional knowledge.

Each of these themes is directly applicable to the CSCD program.

Becoming a Learning Organization

> [A] *Learning Company* is:
> an organisation which facilitates the learning of all of its members and continuously transforms itself.[3]

A learning organization is one that employs specific strategies and tactics to facilitate workforce learning so it may continuously adapt and innovate. Several of the QDDR's goals align with becoming a learning organization, particularly those of harnessing knowledge and technology, promoting innovation, and advancing performance management—which the QDDR anchors to essential components of lessons-learned programs and developing a culture of learning.[4]

The World Bank's International Finance Corporation (IFC) understands the value of having a culture of learning. IFC has four pillars for what it refers to as its *knowledge culture*: practices and processes, incentives, leadership support, and the allocation of resources (see Figure 3.1). Practices and processes are typically informed by policies and procedures and are formal institutional tools. Incentives, which we discuss later in this report, are not as prolific in other organizations as they are in IFC; however, IFC uses its incentive program to "encourage specific behaviors and actions." Leadership support or buy-in is a common positive tool in most of the organizations we interviewed; we explore this further in this section. The allocation of resources is fundamental to successfully establishing and implementing a lessons-learned program.

Figure 3.1
Four Aspects of Knowledge Culture

SOURCE: Sumithra Rajendra and Nadezda Nikiforova, "WBG Knowbel Awards: Building a Knowledge Culture," briefing, April 26, 2016.
RAND *RR1930-3.1*

---

[3] Mike Pedler, Tom Boydell, and John Burgoyne, "Towards the Learning Company," *Management Education and Development,* Vol. 20, No. 1, 1989, p. 2.

[4] U.S. Department of State, 2015, p. 13.

Learning organizations embrace an environment that allows the workforce to admit and openly share lessons learned. Environments that are punitive in the wake of a mistake lead to a culture in which the workforce is less willing to acknowledge mistakes. Schultz offers two models of how errors are viewed: optimistic and pessimistic.[5] Negative feelings are generally associated with making a mistake. This condition is only exacerbated in the work setting in an environment that views errors or being wrong as things to be embarrassed about or as aberrations. Conversely, organizations that find ways to encourage employees to openly share mistakes and to view them more optimistically is better able to promote a culture of learning. Organizations should not make employees feel embarrassed or threatened by mistakes because that promotes "antilearning." Organizations that do so are employing what Argyris refers to as *organizational defenses*, which are considerable barriers to effective lessons-learned practices. This should be avoided in lieu of policies and practices that create an environment of open sharing and communication.

As with most elements of organizational behavior, the leadership sets the tone and directly influences culture. To encourage candor and honesty, leaders need to model behavior that shows that mistakes are tolerated and that learning is encouraged. According to one interviewee, "nobody believes what the leader says—they believe what the leader does, and they follow the leader's example."[6]

Leaders need to communicate that risk and failure are acceptable. Avis and Sharpe acknowledged that

> all members of the organization [must] feel comfortable sharing their mistakes and what they learned from those errors. In order for this to happen, they must be confident that their well-intentioned mistakes will not result in punitive action. . . . If this assurance is not in place, then even the best performance measure will be inaccurate and the overall culture will be one of masking mistakes to avoid punishment—in other words, the complete opposite of a learning environment. In such an atmosphere, it is highly unlikely that lessons will be observed correctly, let alone documented and processed in a manner that will lead to organizational change and improvement.[7]

Leaders can encourage learning by demonstrating the value of information sharing. As one interviewee, a chairman and CEO of a business consulting firm, noted:

> The key to a learning culture is that the quality of a decision has to rise above what the brightest person in the group can offer. The synergy from such information

---

[5] Schulz, 2010, p. 25.

[6] Telephone interview with healthcare industry executive, April 19, 2016.

[7] Avis and Sharpe, 2015, p. 66.

sharing yields decisions that are better than what the smartest person in the group could offer on his own.[8]

As noted in Chapter One, failures and being wrong are vital to learning and positive change.[9] Leaders need to demonstrate by example that the organizational culture is not punitive.[10] Mercy Corps ensures this by conducting panel discussions during which senior leaders talk openly about failures.[11] See box for more information. NASA publishes a newsletter featuring "My Best Mistake" stories, allowing all of NASA to learn and benefit from the constructive mistakes of their colleagues.[12] Robust cultures of learning depend on acceptance and open examination of failures and mistakes as much as they depend on learning from successes.

To encourage candor and honesty, leaders need to model behavior that shows mistakes are tolerated and learning is encouraged. Leaders need to communicate that a zero-defects environment is not an acceptable standard. In our interview with a former CEO of Bloomingdale's, he expressed that the workforce needs to know that taking risk is part of the job and that prudent risks will not be punished if they fail. We define

---

Leaders Play a Key Role in Setting the Organization's Culture

Mercy Corps leadership wrote a new organizational culture statement asserting that Mercy Corps is a risk-taking organization that tolerates failures: "We are not afraid to take risks and will sometimes fail. When that happens we will pick ourselves up, reflect, and try again."

According to Mercy Corps, "Culture, behavior and habits are learned in response to the expectations of leaders. Leaders can set and signal culture through creating a clear vision, messaging and role modelling"—including by holding "teams accountable to learn, change, and show evidence of adaptation and impact" and by enabling and "encourag[ing] team members to admit when there are problems and coach[ing] them to find solutions."

SOURCE: Mercy Corps, "Managing Complexity: Adaptive Management at Mercy Corps," Portland, Oreg., January 2012a, p. 2.

---

[8] Phone interview with the authors, April 12, 2016.

[9] See Shulz, 2010.

[10] Telephone interview with Michael Gould, former CEO of Bloomingdale's, April 8, 2016.

[11] Telephone interview with Mercy Corps Strategy and Learning leader, June 1, 2016.

[12] Interviews with NASA officials by phone, June 13, 2016.

prudent risks as those taken with consideration of risk assessments or awareness of policies, procedures, or plans. In another interview, an executive in the health-care industry stated that

> The first time someone gets punished for their candor is the last time anyone tells the truth.[13]

In our observation, publishing testimonials and storytelling are effective means of demonstrating that no one is penalized for sharing good or bad insights. Published testimonials also make experiences and lessons concrete and relatable.[14] As part of its formal curriculum, Bechtel uses video clips of people from various levels of the workforce relating their experiences. NASA has a program called Pause and Learn, initiated at the Goddard Space Flight Center, that has been institutionalized as a practice across NASA (see box).

Leadership can also build trust by making use of cited values and infusing the corresponding language throughout the company. For example, one large oil company we interviewed reported that innovation and integrity are core corporate values. They use these terms to promote culture by communicating that innovation happens when you learn from mistakes—not when things go well—and that integrity includes acknowledging errors.[15] UPS leadership promotes the idea of "constructive dissatisfaction," which means that employees can be dissatisfied with programs or policies—

NASA's Pause and Learn Program

NASA is strong on case studies and presenting these cases through storytelling. The Pause and Learn Program initiated at the Goddard Space Flight Center has become the standard across NASA and has also been implemented within the Jet Propulsion Laboratory (JPL). JPL has conducted 13 Pause and Learn meetings over the past three years. These meetings are periodic and include the 38 spacecraft project managers telling stories about what went wrong on a project, what they did to fix it, and what they might have done differently. These testimonials are done without slides or displays. The intent of the storytelling is to provide project managers with a forum in which they can admit mistakes and capture lessons learned without fear of attribution or documentation. An added benefit is that the mistakes are not "broadcasted to the world."

---

[13] Telephone interview with health care industry executive, April 19, 2016.

[14] Interview with CLO of hospitality company, McLean, Va., May 5, 2016.

[15] Interview with knowledge advisor at international oil company, May 24, 2016.

the dissatisfaction—but are empowered to make the required change, constructively.[16] USAID uses a slogan tied to its higher core mission of development: "collaborate, learn, adapt, for better development results." These slogans are useful in driving culture and permeating the vision for the lessons-learned program. According to one textbook on organizational theory, "slogans are effective ways of communicating culture because they can be used in a variety of public statements by chief executives."[17] The box illustrates one use of the catchphrase CSI uses to promulgate the value of lessons learned to an organization.

Robust learning cultures also require participation throughout the entire organization, not just the leadership. Lessons-learned programs that are more holistic have broad reach and impact. A large insurance company that we interviewed includes participation in lessons-learned activities in all job descriptions and performance evaluations.[18] In Chapter Five, the section on incentives also discusses how lessons-learned activities can be incorporated into employee performance. A business consulting firm we interviewed reported that its leaders involve a wide range of staff in corporate decisions.[19]

Leaders and lessons-learned actors also could foster trust through how they communicate lessons learned. In a *Harvard Business Review* article, Leon Panetta and his former chief of staff, Jeremy Bash, capture management lessons learned from the hunt for Osama bin Laden.[20] The article aptly addresses one of the challenges the Depart-

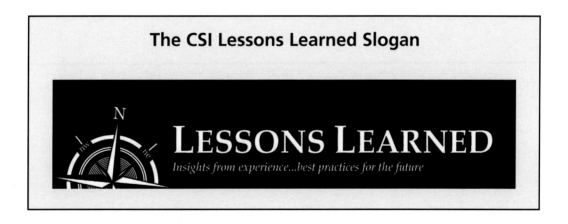

The CSI Lessons Learned Slogan

LESSONS LEARNED

Insights from experience...best practices for the future

---

[16] H. James Harrington, *Business Process Improvement: The Breakthrough Strategy for Total Quality, Productivity, and Competitiveness,* Boston: McGraw-Hill, 1991, p. 248.

[17] Richard L. Daft, *Organization Theory and Design*, 3rd ed., St. Paul, Minn.: West Publishing Company, 1989, p. 510.

[18] Telephone interview with former CEO at a large insurance company, April 11, 2016.

[19] Telephone interview with chairman and CEO of a business consulting firm, April 12, 2016.

[20] Leon E. Panetta and Jeremy Bash, "The Former Head of the CIA on Managing the Hunt for Bin Laden," *Harvard Business Review*, May 2, 2016.

ment of State faces: Every situation is unique. The bin Laden manhunt was unique even in the annals of targeted manhunts—in its duration, its complexity, and the level of human and fiscal resources devoted to it. Yet Panetta manages to extract five management lessons that a corporate vice president for marketing and the director of a multi–billion dollar government agency can both apply.

Although the article highlights successful ideas and initiatives, it also acknowledges mistakes. In this context, Panetta has an advantage in writing about failed efforts because he is no longer in the organization, and the effort was ultimately successful. But in stating bluntly that even he made errors in judgment during the campaign, Panetta serves as a role model for others in the organization. If the boss can admit mistakes, so can everybody.

Panetta also tones down the potential for a lessons-learned retrospective to "shame and blame" by attributing successes to individuals—including not just the President but also individual managers (even if they use pseudonyms)—and attributing shortcomings to unnamed collective groups, such as "a roomful of senior officials," "some inside the Agency," and the generic "we."

Leaders must also foster ways to ensure their actual or perceived views do not dominate the knowledge-sharing activities. This promotes trust and ensures open participation. One interviewee reported asking open questions to ensure that data collection strategies encourage employee participation to prevent everyone from simply agreeing with what the senior person says or feels.[21] A commonly reported tactical approach to doing this is that rank should be ignored when openly transferring knowledge in a trusting environment. The leaders should not allow their status or position in the organization to threaten or intimidate employees from openly sharing:

> Leaders need to demonstrate that "rank doesn't matter when you're problem solving." (Phone interview with a business consulting firm, April 12, 2016)

> "Leaving rank at the door enables candor." (Phone interview with CEO in the health-care industry, April 19, 2016)

> Must take rank off uniform to encourage candor. (Phone interview with CEO in the health-care industry, April 19, 2016)

One expert on organizational culture and change has described how leaders should make themselves open to "being persuaded." Pittampalli wrote that, in "our increasingly complex world, these leaders have realized that the ability to consider emerging evidence and change their minds accordingly provides extraordinary advantages."[22] Describing successful leaders, he further noted that

---

[21] Telephone interview with chairman and CEO of a business consulting firm, April 12, 2016.

[22] Al Pittampalli, "The Best Leaders Allow Themselves to Be Persuaded," *Harvard Business Review,* March 3, 2016.

The winners didn't abide by grand theories of the world, so they were more willing to listen to new information and adjust their predictions accordingly. This willingness to accept the views of others fosters trust and openness and the generation of even better ideas in the organization. Being persuadable comes with two benefits that are vital to any organization: better forecasting of the future and the capacity for accelerated growth.

With respect to better forecasting accuracy, Pittampalli referenced a University of Pennsylvania study by professor Philip Tetlock that tracked 82,361 predictions from over 284 experts and found that accuracy has more to do with how forecasters think than with what they know. Pittampalli reported that the benefit of accelerated growth was reinforced by Swedish psychologist K. Anders Ericsson, who "studied what separates the masters from the mediocre in a wide range of cognitively complex skills (from chess to violin)," and found "that the quality of practice determined performance." Ericsson, according to Pittampalli, observed that quality is based on practice that focuses on weakness, noting that

> Masters were obsessive about identifying and improving on their weaknesses; that means they were able to overcome the natural human bias toward illusory superiority (i.e., the tendency to overestimate our strengths and overlook our faults) by staying open to critical feedback from others.[23]

Pittampalli continued his discussion by advising that leaders should not be persuadable on every issue. This more situational leadership approach considers the contextual circumstances involved in the decision:[24]

> At some point, you have to stop considering new information and opinions, make a decision, and move forward. When time is scarce or the matter at hand isn't very consequential, it's often okay to trust your gut and independently choose a course based on previous convictions. But for higher-stakes decisions, it's important to adopt a more persuadable mindset.[25]

Risk Mitigation

All organizations seek to minimize risk. Over the past decade, lessons-learned programs have come to play pivotal roles in the risk mitigation management. As one interviewee noted, his organization employed "lessons learned [a]s a risk mitigation tool." Yet, unlike other aspects of lessons learned, risk mitigation requires future planning that may be

---

[23] Pittampalli, 2016.

[24] Center for Leadership Studies, "Situational Leadership," web page, undated.

[25] Pittampalli, 2016.

based on pinpointing deficiencies before they have become mistakes or failures.[26] If risk reduction is a goal, the organization must have a strategy to assess weak points that have yet to produce failure and minimize their potential for future repeat.

As demonstrated within the U.S. military, an effective and comprehensive strategy will take an integrated approach to risk that includes lessons learned as a function.[27] This approach helps identify risk areas and related solutions based on the experiences of the organization and its peers to arrive at the least-risky decision more quickly than the organization could otherwise.[28] The causes of risk are certain to differ from one organization to another, however, and are often related to inherent risks of the organization's activities. Brigham and Women's Hospital's Safety Matters program, for example, is designed to promote a culture of safety at the hospital:

> In a culture of safety, staff not only feel safe speaking up about the risks they see, but they are encouraged to do so.[29]

For organizations whose risk potential is difficult to quantify, lessons-learned programs are central to identifying possible threats and challenges as part of the comprehensive organizational management risk plan. The objective of Mercy Corps, for example, is to respond to the humanitarian challenges of crisis and conflict.[30] The environment in which it operates is unpredictable and ambiguous, but lessons learned from a past engagement may still provide a useful template toward identifying and avoiding a similar type of risk in the future.[31] According to Mercy Corps' *Program Management Manual*, "[r]isks can be identified by making use of people who have done similar programs elsewhere, have worked in-country previously and understand the specific challenges, and through lessons learned from other programs."[32] In an unquantifiable field, such as humanitarian work or diplomacy, the qualitative nature of human capital can become the most valuable resource in a comprehensive lessons-learned strategy.[33]

[26] Grabowski and Roberts, 1997.

[27] George Wiklund and Lisa Graf, "Risk, Issues and Lessons Learned: Maximizing Risk Management in the DoD Ground Domain," briefing, Washington, D.C.: U.S. Department of the Army, October 2011.

[28] Telephone interview with lessons-learned coordinator at a global oil company, May 4, 2016.

[29] Karen Fiumara, "Safety Reporting Leads to Safer Systems," Safety Matters blog, Brigham and Women's Hospital, February/March 2016.

[30] Mercy Corps, "Our Mission," web page, undated.

[31] David Garvin, Amy C. Edmondson, and Francesca Gino, "Tool Kit: Is Yours A Learning Organization?" *Harvard Business Review*, March 2008, p. 2.

[32] Mercy Corps, *Program Management Manual*, Portland, Oreg., January 2012b, p. 69.

[33] Karl-Heinz Leitner and Campbell Warden, "Managing and Reporting Knowledge-Based Resources and Processes in Research Organisations: Specifics, Lessons Learned and Perspectives," *Management Accounting Research*, Vol. 15, No. 1, March 2004.

NASA is one entity for which the risk potential is high simply due to the nature of the work and the degree of risk to human life. NASA's work requires extreme precision because it involves highly complex operational environments and processes. Lessons learned are therefore central to organizational culture. JPL, one of the agency's federally funded research and development centers, values the lessons-learned process highly because the laboratory recognizes the potential for error and the cost of repeating an error. According to JPL's standard practice: "Repeated mistakes, or violation of known best practices, pose a risk that is potentially avoidable. . . . [L]essons learned is an effective countermeasure against avoidable risk."[34] To mitigate this risk potential, program managers self-audit their compliance with all documented lessons learned.[35] Regardless of an organization's risk level—high and quantifiable, low and ambiguous, or a combination of these—military services, government agencies, and business enterprises have incorporated lessons learned into effective risk management strategies.

Increasing Efficiency

Government agencies and private industry alike are under constant pressure to improve cost and process efficiency. Within the U.S. military, for example, focus on process efficiency predominates. As the allied joint publication on the conduct of operations states:

> The purpose of a Lessons Learned procedure is to *learn efficiency from experience* and to provide validated justifications for amending the existing way of doing things, *in order to improve performance, both during the course of an operation and for subsequent operations.*[36]

NATO values lessons learned for this purpose, noting that avoiding failure and repeating successes in the military sense means "reduce[d] operational risk, increased cost efficiency, and improved operational effectiveness."[37]

This point highlights that, just as with process pitfalls and failures, past successes can provide the experiences needed to teach lessons-learned practitioners how to repeat what went well. This reduces the need to reinvent the wheel,[38] and leveraging lessons learned makes the most effective use of limited resources.[39] In the private-sector companies we interviewed, production and performance-based companies also

---

[34] Oberhettinger, 2011, slide 2.

[35] Phone interview with GAO manager, April 18, 2016.

[36] NATO, 2011, Para. 0454. Emphasis added.

[37] NATO Joint Analysis and Lessons Learned Centre, "NATO Lessons Learned Process," web page, undated.

[38] CALL, *Establishing a Lessons Learned Program: Observations, Insights, and Lessons*, Fort Leavenworth, Kan., Handbook 11-33, June 2011, pp. 11–33.

[39] Bruce Britton, "Organizational Learning in NGOs: Creating the Motive, Means, and Opportunity," Oxford, U.K.: International NGO Training and Research Centre (INTRAC), Praxis Paper No. 3, March 2015, p. 7.

use process-improvement methodologies to supplement traditional lessons-learned tools and increase the efficiency of production lines. For example, the equipment manufacturing company we interviewed employs Lean Six Sigma, a methodology that aims to amend processes or products to meet customer needs more efficiently, as a substitute for a traditional learning organization.[40] These varying strategic approaches to lessons learned are discussed in detail in Chapter Five. Lessons learned allow an organization to identify clearly what is required, no matter the type of efficiency the company seeks to improve.

Challenges to increasing efficiency may appear when dealing with experiences and processes that are difficult to quantify and resolve with a simple solution. Put more simply, when flexibility is required, efficiency may not necessarily be a priority or even possible. For example, when production speed must be increased, a lessons-learned observation may be able to pinpoint weaknesses in the process chain that requires a simple fix that improves production.[41] Alternatively, in such organizations as Mercy Corps or FEMA, no matter how well planned a relief effort in one state, country, or locality, staff may find a host of unpredictable obstacles in another. An organization's efficiency needs, as well as learning culture, will influence the structure of its lessons-learned program, and vice versa.

Maintaining Institutional Knowledge

Knowledge management, as defined by Gartner Group, is

> a discipline that promotes an integrated approach to identifying, capturing, evaluating, retrieving, and sharing all of an enterprise's information assets. These assets may include databases, documents, policies, procedures, and previously uncaptured expertise and experience in individual workers.[42]

Institutional memory—highly specified, cumulative human knowledge—is a critical function in developing lessons learned, but its priority is easy to overlook when cost and efficiency are pressing. This memory of either an individual or organizational unit of a particular process, functional area, regional area, management, or anything else

---

[40] Telephone interview with Six Sigma expert, April 27, 2016.

[41] Paul S. Adler, Barbara Goldoftas, and David I. Levine, "Flexibility Versus Efficiency? A Case Study of Model Changeovers in the Toyota Production System," *Organization Science*, Vol. 10, No. 1, 1999.

[42] As quoted in Bryant Duhon, "It's All in Our Heads," *Inform*, Vol. 12, No. 8, September 1998. *Knowledge management* is defined many ways; the one used here is most applicable to the context of this study. See also Michael E. D. Koenig, "What Is KM? Knowledge Management Explained," *KM World*, May 4, 2012; Tanvir Hussein and Salim Khan, "Knowledge Management an Instrument for Implementation in Retail Marketing," *MATRIX Academic International Online Journal of Engineering and Technology*, Vol. 3, No. 1, April 2015; and Bahram Meihami and Hussein Meihami, "Knowledge Management a Way to Gain a Competitive Advantage in Firms (Evidence of Manufacturing Companies)," *International Letters of Social and Humanistic Sciences*, Vol. 14, 2014.

can be captured and retained through the knowledge management and integration of lessons learned.[43] Too often, particularly in military and government offices, staff rotate in and out of roles every two or three years. Knowledge may not recorded when personnel leave; even if it is, rarely is read by successors.[44]

Capturing knowledge before it is lost is an important goal for many organizations, and lessons-learned programs can facilitate the process. Through preexit engagement, incentives, or other means, an effective lessons-learned program can house valuable experiences and prevent the lessons from being lost.[45] NASA encourages the sharing of institutional knowledge long before an employee leaves a position through a range of regular initiatives that are available to the entire enterprise online.[46] For example, NASA has found ways to account for the qualitative nature of human capital as it relates to information transfer through informal information exchanges and by capturing anecdotes through a program called JPL Tube (see box).

Private companies are also making the retention of human experience through knowledge management capabilities a priority. At Bechtel, for example, both senior leaders and project managers engage with other companies and foreign governments to negotiate contracts. Those who have been at the company for many years are aware of

---

JPL Tube at NASA

JPL Tube is a formal program inspired by the use of YouTube in less-formal or personal settings. NASA partnered with Microsoft Research to create this program. JPL Tube uses a Microsoft Research product, the Microsoft Audio Video Indexing Service (MAVIS). MAVIS has an upload, transcription, and search capabilities and dissemination restrictions.

JPL Tube allows employees to post footage easily. As part of its appeal, JPL Tube also has an organic element that management does not filter. For example, a retiring employee gave an eight-hour seminar on how to sterilize a spacecraft to avoid introducing earth life to a foreign planet.

NASA management does, however, measure JPL Tube usage. As of our interview with NASA, it had 1,432 active videos; during the last quarter of 2015, there were 325 video views per day at JPL.

---

[43] Jimmy C. Huang and Sue Newell, "Knowledge Integration Processes and Dynamics Within the Context of Cross-Functional Projects," *International Journal of Project Management*, Vol. 21, No. 3, April 2003.

[44] Interview with regional directors at an international bank, Washington, D.C., April 22, 2016.

[45] Britton, 2015, p. 10.

[46] Telephone interview with GAO manager, April 18, 2016.

which companies and countries are difficult to work with and of the history of projects and relationships that have soured as a result.[47] This institutional knowledge is passed down to other rising leaders and project managers via a rigorous management review process for approving contracts. Checklists are available for new program managers to review when starting new projects to avoid some of the mistakes that a seasoned employee would intuitively know to avoid. Indeed, knowledge obtained from engagement remains predominantly peer to peer, but Bechtel's procedures help ensure that, when people leave, their knowledge does not leave with them.[48] Adaptability and a supportive learning environment are key for institutional lessons-learned success and survival over time.[49]

Target Audience

For a lessons-learned program to achieve stated goals, the organization must ask precisely who should be learning and why. This is vital for developing a vision that works in practice because the audience is the recipients of the lessons-learned deliverables. The audience can vary for each lesson and can be individuals or entire business units. Individuals are targeted to change individual behavior that contributes to the overall learning culture. For example, a field worker may be targeted to follow a certain new protocol, a manager or supervisor may be targeted to encourage more staff development, or a senior leader may be targeted to promote a new initiative. On the other hand, a business unit or functional area may be targeted to improve a process, procedure, or standard that exists inside that unit. In both cases there is overlap, given that individuals make up the business unit, so lessons learned often have implications for both the business unit's way of functioning and individuals within a certain unit. Second- and third-order effects of a targeted lesson may be felt throughout the organization, but must have some effect on a specific audience. As one interviewee, a former CEO of a global insurance company, noted

> Companies undertake lessons learned *throughout the organization*, at all levels and in all divisions (but generally closest to the process being evaluated).[50]

Each organization also has different experiences, culture, and goals; a method that is successful in one organization for a similar problem may not necessarily be for another. A supervisor in one unit may be targeted with a lesson about encouraging

---

[47] Interview with Bechtel, Reston, Va., September 23, 2016.

[48] Interview with Bechtel, Reston, Va., September 23, 2016.

[49] Garvin, 2008.

[50] Interview with a former CEO of a global insurance company.

staff development if skills within that unit are underdeveloped. Another supervisor in the same organization might require something else entirely. USMC delivers a specific message to officers at midcareer (around the O-4 level) and above that may not be applicable for Army officers in similar ranks.

Similarly, workers in the field may be targeted with different lessons. One of the private oil companies we interviewed reinforced procedural changes by hanging posters in warehouses that were accessible to field workers. At the CIA, CSI conveys lessons learned outside a classroom setting, focuses on on-the-job learning, and functions independently of the CIA training cell (see box). Conversely, the Department of State houses its lessons-learned center inside its training organization, seeking to incorporate lessons learned into FSI's curriculum. In both cases, training and education are common ways to target individuals in an organization for a desired change. Taking the professional career progression plans of its workforce into account led to CSI's strategy on reaching the workforce in additional ways beyond professional schooling.

In addition to targeting individuals for certain lessons learned, business units can be addressed by looking into establishing or changing certain policies or processes. The Christmas Eve 2013 delivery crisis forced UPS to reconfigure operations at the unit level. A surge in packages led to a delivery bottleneck and national embarrassment as packages expected by Christmas were delivered days later.[51] UPS management incorporated all levels of the company into the lessons-learned exercise and the appropriate business units were executing new procedures nine months later.[52]

The target audience or audiences for a lessons-learned program will depend on these desired outcomes, which will often vary for each functional unit, staff member, and lesson. Once the specific audience is identified, the internal marketing strategy for a particular lesson must then be tailored to meet the different communication styles and mechanisms of that audience.

CSI's 70/20/10 Learning Model

"At CSI, we employ a 70/20/10 learning model. 70% of learning is through on the job training, 20% is peer assisted, and 10% is classroom-based. So, CSI tries to reach people in broader ways than just the 10% of the time they spend in the classroom."

SOURCE: Interview with CSI.

---

[51] Laura Stevens, Serena Ng, and Shelly Banjo, "Behind UPS's Christmas Eve Snafu," *Wall Street Journal*, December 26, 2013.

[52] Interview with Laura Lane, UPS President of Global Public Affairs, Washington, D.C., September 19, 2016.

Governance

> [The office of primary responsibility] is a body consisting of the process owners, the executive leader, and other senior managers, which serves as a strategic oversight body, setting direction and priorities, addressing cross-process issues, and translating enterprise concerns into process issues.[53]

When implementing a lessons-learned program, organizational leaders must determine how it will be governed. Corporate learning literature defines governance as "the strategic oversight required to ensure continued alignment between the company's strategy and the learning organization's mission, and financial oversight to secure adequate resources and funding."[54] One important decision is whether governance should be centralized or decentralized.

Organizations can choose from several combinations of governance structures, and the desired outcomes and target audience will affect how centralized or decentralized a lessons-learned program should be. A centralized governance structure does not mean that the lessons-learned activities must be centralized. The Coast Guard, for example, maintains a centralized lessons-learned capability within the Exercise, Evaluation and Analysis Division, which has centralized control over AARs (which are written by individuals throughout the Coast Guard) and follow-up.[55] A hierarchical, top-down approach is highly effective in the military's structured command environment. Some companies with traditional hierarchical structures may also thrive with a centralized lessons-learned protocol.

Likewise, having a decentralized governance structure does not mean that the organization will not have a top-level office that manages the decentralized lessons-learned activities. An entirely centralized governance structure would likely fail in organizations that have flat management structures, that are driven by individual differences and opinions, or that run field offices in varied environments or with different missions.[56] In such cases, a federated approach with semiautonomous units that report into a centralized body that writes policy and oversees individual units may be most appropriate.

NASA employs one such model; each of its ten field locations specializes in a niche area of expertise and has its own embedded culture. Each location has its own lessons-learned capability with unique means of capturing, evaluating, validating, and

---

[53] Paul Hammer, "What Is Business Process Management?" in Jan vom Brocke and Michael Rosemann, eds., *Handbook on Business Process Management*, 2nd ed., Berlin: Springer-Verlag, 2015, p. 10.

[54] International Institute for Management Development, "Corporate Learning," Insights@IMD, No. 4, Lausanne, Switzerland, May 2011, p. 4.

[55] Interview with U.S. Coast Guard, Washington, D.C., July 6, 2016.

[56] Phone interview with NASA, April 13, 2016. Local research centers may resent a top-down style, so the federated approach is a hybrid of local and centralized control.

disseminating information. Information from all locations, however, feeds into a single NASA lessons-learned information system for organizationwide review, which is facilitated by an agency lessons-learned program manager. Mercy Corps' program also takes a more decentralized approach, deferring lessons-learned processes and decisionmaking to country directors.[57] With this degree of local control, methods for accountability and implementing lessons learned must be made clear.[58]

As a nationwide federal agency, FEMA employs a decentralized model in response to the myriad layers of state and local administrative bureaucracies in which it operates, often during emergencies. A recent initiative involved establishing a component lead for each field office that will manage its own lessons-learned activity according to rules and tools dictated by the central preparedness office. The central office, in turn, works with the field units to assess their successes and failures. This model is effective in that FEMA's lessons learned are geared specifically toward improving field operations that require readily available lessons specific to each geographical area.

This decentralized model with a centralized governance structure may be the best option for the Department of State. The model allows organizations within the department, including different regional or functional offices, to tailor their lessons-learned activities to their unique missions but still provides a centralized body to oversee the coordination of these organizations and to ensure enterprisewide visibility. Whether creating a "learning organization" or a "process office," some unit must serve as the office of primary responsibility for a lessons-learned program. One leading academic on business process improvement (note that the process of conducting lessons learned is in itself a business process) has stated that governance is one of the four prerequisites for an organization to be successful in process management.[59] Specifically Hammer states that "enterprises need a process office that plans and oversees the program as a whole and coordinates process efforts."[60] This office is responsible for creating standards, maintaining tools and guidance, and providing strategic direction for all lessons-learned activities.

No matter how an organization chooses to structure the governance of its lessons-learned program, it must nonetheless decide. Without a formal governance structure, lessons-learned efforts cannot be coordinated across the whole enterprise, losing the positive impact to the organization that the planners sought in the first place.

---

[57] Telephone interview with Mercy Corps senior leader, June 1, 2016.

[58] Telephone interview with lessons-learned leader at an international energy company, May 4, 2016.

[59] The other three are a process-oriented culture, leadership buy-in, and expertise.

[60] Hammer, 2015, p. 10.

CHAPTER FOUR

Who Is Involved? The RAND Lessons-Learned Ecosystem Framework

Having discussed what comprises a lessons-learned program and some program design considerations, as well as best practices for when and why organizations use specific lessons-learned activities, we now turn to who is involved in lessons learned and how various participants relate to one another. This chapter continues the previous discussions on target audience as related to the vision of the organization. To institutionalize the desired effects of the lessons-learned program more holistically into an organization, the entire workforce should be involved in the process. This chapter focuses on both individual actors and organizational entities to set conditions for the discussion of how they interact in lessons-learned programs in Chapter Five.

Actors are individuals or business units charged with implementing and executing the lessons-learned program. Employees can serve as actors tasked with offering lessons learned. Employees can also be the target audience, the recipients of these offerings. Actors and recipients are thus not mutually exclusive groups. For example, depending on an organization's approach to lessons learned, a supervisor can be both an actor and a recipient. This dynamic will be further explored in Chapter Five, which tackles the "how" of lessons-learned programs by discussing strategic approaches to lessons-learned activities.

Working from the literature and expert opinion, we developed a framework that sorts people into three groups essential to the present discussion: members of a chain of command, from lowest level employees to highest level supervisors; members of formal or informal functional units; and members of corporate executive bodies.

In this chapter, we examine the various actors in the lessons-learned process and then discuss the possibilities of target audience and the various supporting infrastructures that can exist depending on the target audience. To understand how the various actors are involved, this chapter first sets the stage by discussing theory on organizational structure to understand how the various components in an organization interact.

After a stage-setting discussion on organizational roles in the lessons-learned process, this chapter introduces the RAND Lessons-Learned Ecosystem Framework, which we developed over the course of this research, to shape best practices for the groups of people involved in lessons-learned efforts. We based this model on organizational theory contributions from the following communities: lessons-learned theory; adult and continuing education; management theory; knowledge management theory; and corporate organizational learning theory. This model identifies key relationships in lessons-learned processes, allowing an organization to identify gaps in its own lessons-learned process where improvements can be made.

The RAND Lessons-Learned Ecosystem Model

We developed a lessons-learned ecosystem model based on Henry Mintzberg's five key components of an organization:[1]

1. strategic apex—top executives of the organization and their direct support staff
2. operational core—the field workers carrying out the core tasks of the organization
3. middle line management—the managers of the operational core, the bridge between the field and the top executives
4. technostructure—technical support (e.g., analysts, engineers, planners)
5. support staff—people who provide indirect service (e.g., human resources, IT).

These components are represented somewhere in each of the organizations that we analyzed for this research, even though specific organizational designs varied greatly. Of particular note is that organizational design often depends heavily on an organization's strategy.[2] There is no one correct way to structure an organization, as long as the top leaders have a consistent vision and strategy and as long as the vision is clearly communicated throughout the entire organization. Any organizational change or improvement has to reflect the vision and be coordinated among all components of the organization.

Our model, depicted in Figure 4.1, identifies the key elements that are essential to an organization's lessons-learned capability: a set of individuals or a business unit; a strategy for governing and driving the lessons-learned activities (which will be extensively explored in Chapter Five); the corporate executive body; intraorganizational enablers; and extraorganizational enablers, which can expand out into the community; and knowledge transfer. With the relationship being further developed next Chapter Five, it is important here to connect knowledge management and the lessons-learned program. Knowledge that has little or nothing to do with lessons learned can be man-

---

[1]   Fred C. Lunenburg, "Organizational Structure: Mintzberg's Framework," *International Journal of Scholarly, Academic, Intellectual Diversity*, Vol. 14, No. 1, 2012.

[2]   Lunenburg, 2012.

Figure 4.1
RAND Lessons-Learned Ecosystem Framework: How Learning Permeates an Organization

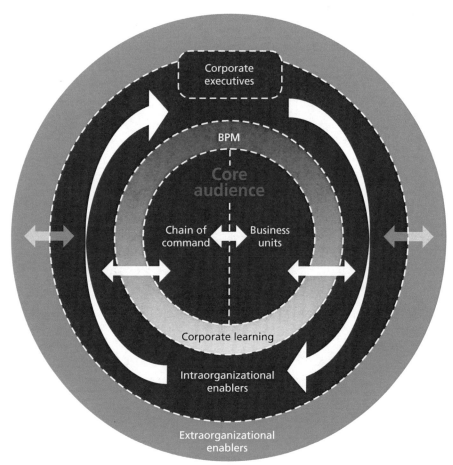

Corporate
executives

BPM

Core
audience

Chain of   Business
command   units

Corporate learning

Intraorganizational
enablers

Extraorganizational
enablers

RAND *RR1930-4.1*

aged or transferred. Conversely, knowledge transfer is a major enabler of any effective lessons-learned program.

No single organizational cell or function can be responsible for all the steps in the lessons-learned process. Consequently, an organization must have an infrastructure that accommodates an integrated approach to both generating and incorporating new ideas and innovating from lessons-learned experiences. As discussed in Chapter Three, either a centralized or a decentralized approach can be effective, but the key is creating a governance structure to oversee the lessons-learned capability. This governance provides formal infrastructure that in turn provides legitimacy and, therefore, resources to the lessons-learned program.

Knowledge Transfer

Knowledge transfer is a key enabler to effectively conducting lessons-learned activities. The arrows in Figure 4.1 depict the flow of knowledge throughout the organizational components. Specifically, these arrows represent knowledge transfers, which are defined as "the process through which one unit (e.g., group, department, or division) is affected by the experience of another."[3] Vision and strategic initiatives flow from the corporate executives to the organization. Business units and supervisors collect lessons and often implement new processes and procedures to match these strategic initiatives. These new initiatives cannot be implemented fully without the support of enablers, such as training, information services, and human capital planning, including extra-organizational enablers. Results are then communicated to the corporate executives.

As stated earlier, while not all knowledge transfers necessarily relate to a lessons-learned program, they can still contribute to the overall learning culture of an organization. Human interactions, formal or informal, contribute to adult learning.[4] The relationships that adults form both inside and outside the workplace allow informal lessons to be learned. Coworkers not bound by structure can exchange information, as can other communities of practice involving similarly skilled workers. Communities of practice and outside professional organizations can also inform an individual's work.

Finally, the middle circle in Figure 4.1 depicts the spectrum of tools and approaches an organization can use to implement lessons learned, from BPM to corporate learning, as discussed further in Chapters Five and Six.[5]

The Core: The Target Audience

At the core of the model is the target audience, which is best described as who will benefit from the lessons-learned activity. As indicated in the vision discussion of Chapter Three, the decision about who is to be targeted drives the formal and informal infrastructure put in place and influences the centralized and decentralized lessons-learned activity in the organization. The core is divided into two primary groups: (1) the workforce that exists inside a chain of command consisting of individual employees and/or supervisors and management and (2) functional business units that can be either formal or informal.

Although individuals in the workforce may be targeted for a variety of reasons, such practices as training, performance reviews, and incentives are tools that target individuals. Generally, these tools promote the desired organizational culture. On the

---

[3] Linda Argote and Paul Ingram, "Knowledge Transfer: A Basis for Competitive Advantage in Firms," *Organizational Behavior and Human Decision Processes*, Vol. 82, No. 1, 2000.

[4] Sharan B. Merriam and Phyllis M. Cunningham, eds., *Handbook of Adult and Continuing Education*, San Francisco: Jossey-Bass Publishers, 1989.

[5] BPM is an operations management subfield. BPM focuses on improving the operational processes that allow an organization to conduct its business (see Hammer, 2015). Corporate learning is a holistic approach to capturing and disseminating best practices through an institutionalized culture of learning.

other hand, business units are targeted through process improvement and policy or procedure changes to implement lessons learned. From an organizational perspective, business units contain the operational core, middle managers, and sometimes technical support. These units align with the strategic apex of an organization through policy and procedure to maintain organizational standards.

The Corporate Executive Body

Corporate executives provide the vision for and priorities of an organization, and a lessons-learned program must fit in with these. However, senior leaders then must support the lessons-learned process by emphasizing its importance to the rest of the organization: "It is an absolute necessity that the leadership of the organization visibly support the process and participate in influencing a learning culture through both policy and resourcing."[6] Yet, senior leader buy-in is a common major barrier to successful lessons learned.[7]

Indeed, senior leaders with limited time and pressing concerns might find it difficult to make lessons-learned buy-in strategies a priority. Internal promotion of lessons-learned processes to senior leaders by those already committed serves to break this primary barrier. This can be achieved through senior-to-senior communication in conferences and meetings, collaboration in task forces or virtual teaming, and regularly scheduled time committed to communication on this topic alone.[8] At UPS, for example, executives lead business resource groups—innovation groups of employees with particular personal affinities or identities. The groups are guaranteed executive attention and interest because the executives lead the groups.

Senior executives are also able to achieve buy-in across the organization by encouraging an open environment in which "constant questioning, critical thinking" are core functions.[9] By creating the type of forum in which all ranks of the organization are welcome to provide input on lessons learned, seniors are able to establish an informal internal marketing arena that encourages participation from all staff levels.

An example of including all ranks in lessons-learned discussions is NASA's "program manager challenge," which encourages leaders to discuss lessons learned in a live web telecast that is available to a range of NASA staff. Another forum, known as Masters with Masters, brings in senior management from across NASA for live webcasts via

---

[6]   Avis and Sharpe, p. 65.

[7]   Many interviewees attested to this fact. Additionally, Milton, 2010, p. 129, lists leadership buy-in as the second biggest barrier to lessons learned, and Hammer, 2015, p. 10, lists senior leadership buy-in as the most difficult of the BPM perquisites to overcome.

[8]   Jongwoo Jeon, "Success Factors for a Lessons-Learned System in a Construction Organization," *Cost Engineering*, Vol. 51, No. 5, May 2009, p. 13.

[9]   Marlin, 2008.

YouTube.[10] Members of the junior staff are more likely to buy into the process if they consider themselves part of it. In addition to creating commitment among seniors by seniors, simply opening the space for input from other functional areas creates more potential for buy-in than if this input were discouraged.

Intraorganizational Enablers

Most organizations have infrastructure dedicated to improving the organization, which is one of the main motivators for establishing lessons learned and knowledge transfer capabilities. These functions manifest in a variety of ways; some organizations have CLOs, others have integrated product teams or process improvement divisions. Some organizations have training inside a human resources cell, and others pull training out as a separate entity.

Some organizations have their chief information officers (CIOs) manage the knowledge-sharing IT platform, while others assign their CLOs the responsibility of knowledge-sharing processes. Several models exist, but all can be successful as long as the key tasks necessary for lessons learned are represented somewhere in the organization. Our analysis determined that the organization should have infrastructure for the following key tasks:

- knowledge management (includes knowledge sharing and dissemination)
- talent management (including skill identifiers)
- collection of knowledge
- writing policy and doctrine
- maintaining an incentives structure
- developing innovative solutions
- teaching and training.

So, although one office may ultimately be responsible for lessons learned, this office must integrate with other offices to ensure that the lessons learned are successfully implemented. Chapter Five discusses in detail the infrastructure that supports lessons learned and that is responsible for the key tasks just listed. In the next subsection, we look more specifically at the people who conduct lessons-learned studies.

Who Conducts Lesson Learned Studies?

As discussed in the Chapter Two, the vision for a lessons-learned program will affect how it is set up and who is involved in it. For example, if the goal of the program is to keep field workers informed, a federated, decentralized model may be more appropriate. In this model, a lessons-learned expert may be assigned to work with employees

---

[10] NASA Office of the Chief Knowledge Officer, "Sharing Masters with Masters," web page, undated.

in the field. The number of people involved such a program and where they sit in the organization can vary.

Generally, in a centralized model, a centralized, dedicated lessons-learned staff will oversee and conduct lessons learned. A dedicated team is insulated from other organizational pressures so it can focus on learning.[11] One of the interviewed oil companies has 3,000 global employees but a centralized lessons-learned team of two people. They are able to conduct 40 lessons-learned sessions and three "lookbacks" per quarter. Having a centralized dedicated staff also allows assessments to be corporatewide efforts incorporating many stakeholders and mitigating any sense of intrusive scrutiny or blame the team being assessed might feel.[12]

Some centralized lessons-learned programs serve as overseers and integrators, while actual business units conduct lessons-learned reviews. For example, one oil company that we interviewed has a centralized lessons-learned staff of six located in its corporate headquarters who curate the lessons learned that are posted by business units; the business units are responsible for conducting lessons-learned reviews. The lessons-learned staff focuses on how to extrapolate from the learning experiences for application to multiple sections of the company.[13]

When the lessons are more tactical or operational, the lessons-learned program performs better when those doing the work take ownership of the program. This is true in FEMA and at an interviewed hotel chain, which advised undertaking the lessons-learned program at the lowest possible level, the one closest to where the work is done.[14] Even in more decentralized models, instead of teams of lessons-learned experts, individual lessons-learned experts can reside throughout the organization, in various business units. In companies that emphasize the use of process-improvement techniques, such as Six Sigma, each individual in the company is trained in basic Six Sigma methods with experts (called "Master Black Belts") strategically placed throughout the organization.

Extraorganizational Enablers

External enablers extend as far outward from the organization as the macroenvironment in which the organization exists and are generally described as any knowledge-producing source that exists outside the organization.[15] Specifically, external enablers are

---

[11] Telephone interview with Mercy Corps Strategy and Learning leader, interview with the authors by phone, June 1, 2016.

[12] Telephone interview with lessons learned coordinator at a global oil company, May 4, 2016.

[13] Interview with knowledge advisor at global oil company, May 24, 2016.

[14] Interview with CLO of hospitality company, McLean, Va., May 5, 2016.

[15] Merriam and Cunningham, 1989, p. 252.

knowledge resources existing outside the organization which could be used to enhance the performance of the organization and they include explicit elements like publications, as well as tacit elements found in communities of practice that span beyond the organization's borders.[16]

This could be as simple as two corporate learning officers meeting at a roundtable event and making plans to share ideas later, which happened at one of the roundtables that informed this report. Another example is the use of contractor organizations to facilitate lessons learned and knowledge management activities. One interviewee described to us how Coca-Cola uses extraorganizational enablers (see box).

Coca-Cola's Project Last Mile Partners with Extraorganizational Enabler

"Millions of people in hard-to-reach parts of Africa lack adequate access to life-saving medicines and medical supplies. Despite valiant efforts by donors and local governments, heartbreakingly, these drugs and supplies are often within reach but still don't make it to those in need or spoil before they do.

The Coca-Cola system, which routinely delivers its products to some of the remotest parts on earth, saw an opportunity to use its logistics, supply chain, and marketing expertise to help remedy this situation. Together with USAID, the Bill and Melinda Gates Foundation, and the Global Fund, they created an initiative called 'Project Last Mile,' which leverages best practices from Coca-Cola's business model to address bottlenecks in medicine distribution. This is done through a facilitated process, where Coca-Cola adapts, customizes, and shares its business and distribution practices.

Recognizing that monitoring and measuring their results would be critical to scaling the project and maximizing its impact, the Project Last Mile team invited The Yale University Global Health Leadership Institute (GHLI) to join the initiative. In addition to identifying improvement opportunities, GHLI provides an important academic link, allowing the team to share its 'lessons learned' with the broader development community."

The Coca-Cola Company

SOURCE: Interview with Coca-Cola executives.

---

[16] Alan Frost, "Introducing Organizational Knowledge," KMT: Knowledge Management Tools website, 2010.

CHAPTER FIVE

How Do Organizations Approach Lessons-Learned Program Activities?

How organizations conduct lessons learned is closely connected with who is involved in lessons learned, as was discussed in the previous chapter, using the RAND ecosystem model as the framework. This chapter expands on the notion of enablers introduced in Chapter Four by first discussing the various strategic approaches to lessons learned—corporate learning, BPM, and hybridization—and then exploring the associated enabling infrastructures. Enabling infrastructures include human capital management, resourcing, marketing, incentives, and knowledge management. RAND's Lessons-Learned Ecosystem Framework and associated organizational theory will be referenced throughout the chapter (see Figure 4.1).

Strategic Approaches to Lessons Learned

Organizations can use three primary strategic approaches exist to drive lessons-learned programs and/or the transfer of knowledge: corporate learning, BPM, and a hybrid of the two. Corporate learning is a holistic approach to capturing and disseminating best practices through an institutionalized culture of learning. BPM is "a comprehensive system for managing and transforming organizational operations."[1] Both strategies have the same goal. They stress the need for leadership buy-in and a work culture that is adaptive and innovative. But they also have two key differences. First, corporate learning tends to emphasize knowledge and knowledge sharing, while BPM tends to emphasize process and process improvement. Second, corporate learning tends to advocate for a learning organization that functions outside normal business units and the chain of command, while BPM emphasizes that corporate change should be managed by leaders within main business units.

It is important to note that corporate learning cannot fully integrate lessons learned without paying attention to processes, and BPM cannot improve processes

---

[1] Hammer, 2015, p. 3.

without knowledge sharing. Furthermore, even in a corporate learning model, supervisors and managers within business units have to implement the lessons learned. Neither strategy is more optimal. Functional business units in the same organization may require different approaches based on their outputs and mission. The establishment of dedicated infrastructure to execute key tasks is what matters the most.

Corporate Learning

Corporate learning is defined as "the capacity of an organization to acquire, apply and share knowledge for the purpose of exploring new solutions and exploiting them to improve efficiency and competitive advantage."[2] According to Shlomo Ben-Hur, a well-published professor of corporate learning, there are three main stages of corporate learning:[3]

1. **Acquiring knowledge.** An organization must first figure out what needs to be learned, both from internal feedback mechanisms and as a way to test external demands on the system.
2. **Applying knowledge.** Next, an organization must test this new knowledge and apply it to improve the situation.
3. **Sharing knowledge.** The new knowledge must then be shared throughout the organization.

Ben-Hur stressed that creating a learning culture within an organization is imperative for success. Cultural enablers include creating an organization in which people feel safe to suggest new ideas, have autonomy to initiate new ideas, and are rewarded for new ideas. Incentives are important for creating a learning culture. It is also important, according to Ben-Hur, to be cognizant of obstacles a learning organization typically faces. Organizational positioning and aligning with corporate strategy are two of the most difficult areas that can render learning organizations ineffective.[4]

The CIA's lessons-learned model adheres closely to the corporate learning model. CSI is organizationally positioned outside the CIA's core business lines. This fosters intellectual independence and gives CSI autonomy to write fair, balanced, and objective products. The intrinsic challenge of this model is making sure that lessons are properly disseminated or easily accessible and that the lessons are integrated into the CIA's daily operations. To address this, CSI must actively engage with business units to maintain relevancy. The Department of State is also currently set up to pursue the corporate learning model. Its lessons-learned center, CSCD, is organizationally posi-

---

[2]   International Institute for Management Development, 2011.

[3]   Illustrated in International Institute for Management Development, 2011.

[4]   International Institute for Management Development, 2011; Shlomo Ben-Hur, Bernard Jaworski, and David Gray, "Aligning Corporate Learning with Strategy," *MIT Sloan Management Review*, September 1, 2015.

tioned within FSI. The current reach of CSCD is limited to foreign service officers who take training courses at FSI, but overcoming organizational positioning and becoming more integrated with daily diplomatic operations could make CSCD an effective tool for disseminating lessons across the department.

Business Process Management

Another strategy to improve organizational performance is BPM, which focuses on improving end-to-end business processes. BPM is a corporate-level strategic approach to managing business processes that employs and synthesizes various management trends and tools to achieve its goals. Some tools of BPM include Lean Six Sigma, an approach to management that focuses on eliminating waste or unnecessary steps in any process while ensuring quality remains high, and business process reengineering (BPR), which is improving a process by completely redesigning it from beginning to end.[5] Harmon explains this relationship of business process trends through the pyramid depicted in Figure 5.1, which segments an organization into enterprise, business process, and implementation levels.

Figure 5.1
Business Process Trends Pyramid

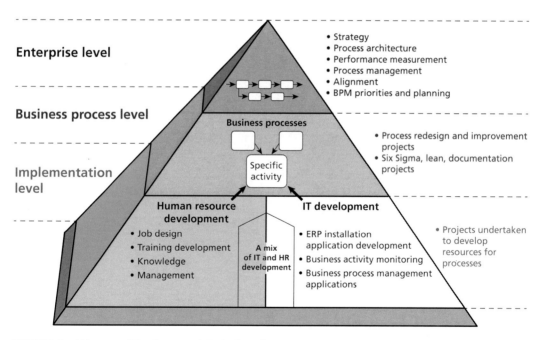

SOURCE: Paul Harmon, "The Scope and Evolution of Business Process Management," in vom Brocke and Rosemann, 2015, p. 54. Used with permission.
NOTE: ERP = enterprise resource planning.
RAND *RR1930-5.1*

---

[5]   Definitions can be found on the Investopedia website.

At the business process level, business management experts have codified three major business process traditions: quality control, management, and information technology.[6] Each tradition provides a different approach to improving business processes.

The *quality control* approach to process improvement focuses on the quality and the production of products. It looks at making incremental changes to existing processes for smaller cost savings. Such quality control approaches are continuous process improvement, Lean Six Sigma, and Total Quality Management.

The *management* approach emphasizes organizing and managing employees to achieve corporate goals. The focus is on the overall performance of the firm and aligning business processes with corporate strategy. One example of a management tool is the agile method, which focuses on methods being adaptive and people oriented. Some consider BPR to be a management approach because it motivates senior executives to completely rethink their business strategies. GE, one of our interviewees, shared its experiences with the use of FastWorks, a type of agile methodology (see box).

Finally, the *information technology* tradition focuses on the use of computers and software applications to automate work processes. Many government organizations successfully reduce the time and number of steps required in a process through IT automation. Since BPR requires processes to be redesigned for maximum efficiency, BPR is also considered an IT solution.

General Electric's Use of FastWorks

FastWorks is GE's take on the "lean startup" methodology of Eric Ries, which focuses on developing new products quickly to learn quickly as well. FastWorks gives small, cross-functional teams high levels of autonomy so they can take charge of the development process. Each team interacts with customers, designers, suppliers, and other development-process stakeholders to test constantly until the best product is created—then the team begins again, developing the next product. This differed from their previous product timeline, redesigning every five years and keeping new products well hidden. GE realized that speed would be to its competitive advantage and made management and procedural changes to allow these small teams to experiment and learn as quickly as possible.

The results for one group, GE Appliance, were positive. FastWorks allowed the group to halve costs, double development speed, and sell at "over two times the normal sales rate."

SOURCE: Brad Power, "How GE Applies Lean Startup Practices," *Harvard Business Review*, April 23, 2014.

---

[6] Harmon, 2015, p. 54.

The tool of applying lessons learned tends to be associated with corporate learning, so it is important to explore how lessons learned fit into BPM. Regardless of the management traditions or tools used, process improvement techniques generally follow a cycle similar to that for lessons learned, as described in Chapter Two. "A typical process improvement initiative will undergo the following steps":

- map the target business process
- identify and remove wastes
- identify problems
- prioritize problems
- identify problem root causes and corrective measures
- analyze alternatives
- redesign the process.[7]

To this, we have added one more item: implementation. The chapter the list was taken from specifically states that it is looking at the "description and exemplification of these techniques," acknowledging that the authors left out measuring performance and implementation strategies.

These steps similarly match the steps of the lessons-learned cycle: action and analysis, lesson identified, and implementation. As in the lessons-learned cycle, process improvement steps involve collecting data on the area that needs change, assessing which changes are most important, and determining what actions can be taken to rectify the problem. Additionally, implementation requires much oversight and uses performance measures to determine the success of a redesigned process.

Two of the interviewed organizations followed a more traditional BPM framework. One specifically used the process improvement tool known as Lean Six Sigma, mentioned previously, in the quality-control tradition. This equipment manufacturing company trained all its employees in basic Lean Six Sigma principles, making them green belts. Additionally, each project manager had an advanced Lean Six Sigma employee, known as a black belt, reporting directly to him or her and supervising the work, ensuring that Lean Six Sigma methodologies were used to reduce costs where possible. Organizational change happens inside the business units with managers and supervisors in charge of capturing and implementing new procedures.

One of the interviewed global oil companies had a strategic management framework that it applied to the entire company. In this framework, all levels of employees strictly follow a set of standards that are well distributed and well known. To counter a heavy reliance on published standards that can sometimes stifle agility, this company boasts a wide array of ways to change the base standards. Employees are encouraged to suggest changes; if any employee discovers an innovative way to make improvements,

---

[7]   vom Brocke and Rosemann, 2015, pp. 129–130.

there are procedures to allow. Suggestions are carefully reviewed up the chain of command and, when appropriate, are used to update the global standards.

We also interviewed a large multinational technology company that intentionally did not institute any form of centralized corporate lessons-learned structure. The individual said that they could not find

> anyone within the corporate hierarchy who wears the responsibility for "lessons learned" because, since [my company] prides itself on having a flat organization, this responsibility is managed at very low organizational levels and is focused on specific areas of concern.

On further analysis of this organization, it became clear that lessons learned is one of the core business practices at the business unit level rather than being undertaken as a higher-level review or as an after-the-fact postmortem. This organization is a global leader in innovation, has dedicated positions throughout, and uses this infrastructure and its managers to drive change.

Hybridization

There is no right or wrong way to set up the processes and procedures for conducting lessons learned. Most organizations have multiple integrated approaches throughout, both within business units and in support functions. These approaches use both BPM tools and corporate learning practices. Many of the organizations we interviewed used a hybrid model to implement lessons learned. In the Army, for example, CALL collects and disseminates best practices. A related organization, the Army Office of Business Transformation, oversees and writes policy for improving the Army's business mission area through BPM tools, such as continuous process improvement and BPR.

The challenge is integrating these approaches. On a smaller scale, the global construction company Bechtel has multiple approaches to organizational improvement. It employs a CLO who focuses on building curriculum for Bechtel training courses, while also employing a CIO to manage a knowledge-sharing platform. The CIO's platform houses both the Lean Six Sigma program and the lessons-learned program. The goal is to create a system that allows lessons learned to permeate the entire organization and thereby involve the entire workforce. Successful implementation of organizational change requires the whole organization to buy in to the change.

One interviewee, a retired general from the U.S. special forces, cautioned against creating an isolated lessons-learned office. His experiences indicate that an organization risks forgetting or ignoring a lessons-learned office that operates apart from the rest. If employees' daily work does not involve lessons learned, they might be less willing to devote valuable time to what they might consider an extra or unnecessary duty. Isolating the lessons-learned office, according to the general, creates a bureaucratic membrane that might prevent others in the organization from engaging.

Associated Enabling Structures

The organizational theories that inform this discussion—lessons-learned theory, adult and continuing education, management theory, knowledge management theory, and corporate and/or organizational learning theory—all identify organizational enabling elements that are necessary prerequisites to implementing change successfully. Drawing from this literature and interviews with practitioners, we identified key supporting infrastructures necessary for successful lessons learned.

Human Capital Management

The organizations we interviewed overwhelmingly acknowledge the importance of human capital management for successful lessons-learned programs; all military organizations and nearly all government and private sector organizations noted the need for developed talent management and training programs.[8] All lessons-learned efforts must include both experienced lessons-learned practitioners and SMEs. Ways to integrate these roles vary. In some cases, such as in the military, lessons-learned staff members will include generalists and lessons-learned practitioners, with SMEs drawn from business units or outside sources. Other organizations, such as one of the large oil companies we interviewed, trained SMEs how to lead and facilitate candid discussions and conduct lessons-learned reviews in their functional areas of expertise.[9]

Organizations with limited resources have found ways to optimize staffing, such as using a matrix and shortening the lessons-learned life cycle by including end-point recipients of the implementation outputs as collectors (curriculum writers, doctrine writers, trainers). For example, both the Navy and the Army use doctrine writers who are experts in their respective fields to participate in collection efforts.[10]

Training staff to be lessons-learned practitioners is also important for developing a successful lessons-learned program. Inside the military, training is a key aspect of improving the quality of data culled from external collections and of increasing the organization's credibility externally. The USMC and USAF provided at least three days of formalized lessons-learned training to anyone who would be conducting lessons-learned studies. The more people trained in lessons-learned skills, the more widespread is the commitment to undertaking lessons-learned efforts and changing business practices based on past performance.[11] Training is also key after lessons have been identi-

---

[8] See Appendix B for more detail.

[9] Interview with knowledge advisor at international oil company, May 24, 2016.

[10] Interview with Wayne Baugh, Chief, Combat Training Centers and Lessons Learned Branch, Training and Doctrine Integration Directorate, Combined Arms Support Command, Fort Lee, Va., May 2, 2016.

[11] Telephone interview with Robert Lefton, chairman and CEO, Psychological Associates, April 12, 2016.

fied and when business process change needs to be implemented. A hotel chain that we interviewed updates its training to reflect all business process changes.[12]

Resourcing

One of the biggest challenges for lessons-learned programs is obtaining a secure source of funding. If an organization does not see lessons learned as a priority, it will not emphasize or resource lessons-learned activities. To become a priority, lessons-learned programs have to prove credibility by providing some sort of return on investment. This return may not be financial but has to be tangible enough for many in the organization to buy in to the program. Staff must see learning as contributing to the organization's mission; as a legitimate, value-added activity; and as benefitting, rather than harming, their careers.[13]

Organizational position is a key factor in securing funds as well. According to one expert in organizational development and organizational learning, it is vital to "[e]nsure there is visible sponsorship for the process and the right resources available to carry out, facilitate, document, and present the results."[14] A cell that is sponsored by key senior executives is more likely to have secure resources. The Marine Corps Center for Lessons Learned is embedded inside the training command but is not essential to the daily conduct or function of USMC training. Because of this, its resources may be more at risk for funding reductions than the training command's more strategically vital training programs. On the other hand, USAF's lessons-learned organization is organizationally positioned in USAF Headquarters (A9) and is tied to highest level USAF decisionmaking bodies, suggesting that its resources are far more secure.

To offset resourcing problems, some organizations decrease the scope of their lessons-learned programs. The USMC, for example, chooses to target only senior leadership as the audience for its lessons-learned program. The program does not reach the entire organization, but its products are of high quality and value to senior leaders. CCO uses temporary unpaid interns to write and publish a magazine called *PRISM*, which reports on the center's activities and research.[15]

We found another related example of the significance of organizational positioning in the CIA. CSI was at one time part of the organization's training center. This alignment did not work well because the work that CSI does on a daily basis was fundamentally different from that of its training counterparts, so it was often difficult for the training center to map CSI's work back to organizational goals and objectives. The

---

[12] Interview with CLO of hospitality company, McLean, Va., May 5, 2016.

[13] Britton, 2015, p. 16.

[14] Ernst and Young, "Profiting from Experience: Realizing Tangible Business Value from Programme Investment with Lessons Learned Reviews," pamphlet, London, 2007, p. 4.

[15] Interview with CCO, Washington, D.C., August 19, 2016.

ensuing change resulted in an alignment scheme that placed CSI outside the core business lines of the CIA (see box)—which CSI and some other interviewed organizations found valuable because employees were free to identify and articulate potential needed changes.

Marketing

A lessons-learned process, even if well planned and structured, will find it difficult to survive if internal stakeholders are not convinced that the process is a key function. Moreover, an organization's human capital is more inclined to engage in earnest if the process is seen as worthwhile to both the individual and the broader organization. A range of internal education and marketing strategies can and should be employed to obtain buy-in from the top down and the bottom up. The purpose of marketing inside an organization is "to inform, to stimulate, and to meet the needs of the clients."[16] The breadth, depth, and style are contingent on the scope and resources available to advertise to the organization.

Marketing can have a formal or informal appearance and can be disseminated either actively or passively, online or offline.[17] Formal and active means, such as with conferences or targeted newsletter distribution, are one way to advertise, but more casual advertising can be done on a regular basis. For example, a hotel chain interviewed actively disseminated lessons learned at daily shift kickoff meetings.[18] Alternatively, all staff levels can and do market by word of mouth on informal, unplanned basis, such as through ad hoc lunches or open forums where open communication regarding problems and challenges is encouraged.[19] Buy-in is effectively earned through consultation.

The Center for the Study of Intelligence— Separate and Active

CSI is separate from the CIA's core business lines. This organizational positioning enables it to write fair, balanced, and objective products. CSI reports that it has to engage business units to be relevant and for people to know what it is and what it does. To overcome this, CSI builds awareness of its capability through marketing, outreach, and a poster campaign that aligns CSI's learning goals with the overall goals of the CIA.

---

[16] Merriam and Cunningham, 1989, p. 250.

[17] Jeon, 2009, pp. 51–55.

[18] Interview with CLO of hospitality company, McLean, Va., May 5, 2016.

[19] Susan McIntyre, *Utilizing Evidence-Based Lessons Learned for Enhanced Organizational Innovation and Change*, Hershey, Pa.: IGI Global, 2015.

Passively communicating lessons learned is also effective, such as through an organizational website or e-bulletin, but success typically depends on employee access or bulletin visibility. The Department of State's use of the roundtables in connection with this research is also a great example of marketing.

Targeted messaging with the aim of encouraging participation and acceptance of lessons learned as an organizational process can complement passive distribution means. The messaging will nonetheless differ depending on the audience. For example, senior leadership will respond to certain marketing stimuli differently from the way junior staff will. By identifying the recipient, the means of delivering the message can be tailored to maximize buy-in potential within a certain group. Mercy Corps, for example, disseminates lessons-learned information in an online newsletter with no target audience that virtually all of its staff accesses. [20] By contrast, NASA's well-developed lessons-learned portal emails lessons-learned studies to targeted individuals through its built-in subscription feature.[21] CSI has a poster that is distributed throughout CIA and Intelligence Community work locations that advertises the existence of the capability, demonstrates the importance of the capability, and that prescribes laypersons should conduct lessons learned (see Figure 5.2).

NASA's advantage is that its knowledge database is widely accessed at all staff levels. If a knowledge database is operational, staff must be informed that it is ready for use, be trained to use it, and be instructed on how to integrate its content into their work processes.[22] Arranging informational seminars and symposia that identify the ways that lessons learned can benefit the team and organization at large can serve the twofold aim of securing buy-in and educating those who will be part of the framework.[23]

Incentives

In addition to marketing, incentive programs are key to attracting stakeholder buy-in across an organization. One method is to include lessons-learned processes in job descriptions. At Mercy Corps, for example, all job descriptions state that all staff are expected to devote 5 percent of their time to formal and/or informal professional learning activities. This type of incentive not only prioritizes lessons learned but also legitimizes workers spending time on learning activities that otherwise fall outside their normal responsibilities. Bechtel has a learning culture that is so ingrained that employees who are asked to facilitate training courses readily agree because it is viewed as an essential part of the job and a great honor to be asked.

---

[20] Mercy Corps, 2015.

[21] Oberhettinger, 2011, slide 6.

[22] Telephone interview with David Oberhettinger, Office of the Chief Engineer, Jet Propulsion Laboratory, California Institute of Technology, April 13, 2016.

[23] McIntyre, 2015.

Figure 5.2
CSI Lessons-Learned Marketing

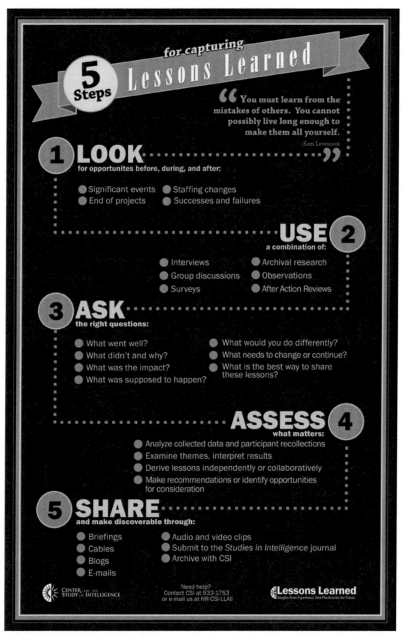

SOURCE: Poster from CSI.

Providing rewards is also another successful incentive for lessons learned. Organizations can reward employees with bonuses or promotion for participating in lessons-learned activities. This is achieved by including lessons learned in the performance review process. Additionally, awarding prizes or giving recognition for innovative initiatives can improve buy-in. Both interviewed oil companies present annual senior leadership awards with high visibility. Recognition alone can be a powerful incentive, especially in resource-constrained organizations. NASA reports that it recognizes employees through the JPL newsletter, a publication with secure funding, making this type of recognition nearly costless.

Several organizations we interviewed or investigated have advanced incentive programs. This drives culture and generates great ideas for improving the company. A large oil company we interviewed has an award program for those who come forward to identify mistakes or areas of improvement, demonstrating integrity and responsibility. Financial values are attached to these lessons—how much money they will save the company in the future. This named award program in the oil company focuses on the "future value of fixing a problem and not on the past cost of having made the mistake." Another oil company we interviewed also has an incentive program that relies on awards and bonuses and has a competition-based incentive. The company conducts the competition annually and typically has more than 600 applicants—which represents 600 pulls of innovative ideas from the workforce. Encouraging employees to be reflective and to innovate to improve is part of the bedrock of a strong culture of learning.

We also encountered the incentive program of the knowledge management program at the World Bank's IFC. IFC postulates that a successful incentive program will

- encourage specific behaviors
- align with business goals
- be time-bound
- have measurable outcomes
- offer rewards that are meaningful to staff.[24]

IFC has structured its KNOWbel Award program to be a competition-based incentive program that employees use to generate new ideas. It is one of several examples we noted of means organizations use to pull creative ideas and capture lessons learned from the workforce. The KNOWbel program is structured as shown in Figure 5.3. To evaluate KNOWbel submissions, IFC created a system that uses the acronym *SCORE*, as described in Figure 5.4. In addition, IFC has standardized submission forms to evaluate the entries that include lessons learned during the effort, as shown in Table 5.1. The KNOWbel incentive program's detailed structure and scoring system, as shown in these figures, is yet another example of an incentive program the Department of State may want to consider.

---

[24] Rajendra and Nikiforova, 2016.

Figure 5.3
Structure of KNOWbel Awards Program

Aligned to business outcomes	Visibility	Measurable

• Categories:
  – Client solutions
  – Increased efficiency
  – Learning from the past
• Strategic themes
	• Leadership meetings
	• Staff recognized throughout the year
	• Enabled staff to set up brown-bag lunches
		• Clear criteria in submission form
		• Process, results, and lessons learned

☆☆☆ **Reward** ☆☆☆
• Visibility of project by leadership
• Peer recognition
• Nominal monetary award and a plaque

SOURCE: Adapted from Rajendra and Nikiforova, 2016.
RAND *RR1930-5.3*

Figure 5.4
IFC SCORE System

SCORE
An easy way to make knowledge a part of your work

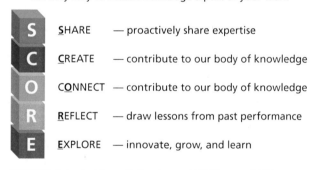

S̲HARE — proactively share expertise

C̲REATE — contribute to our body of knowledge

C̲O̲NNECT — contribute to our body of knowledge

R̲EFLECT — draw lessons from past performance

E̲XPLORE — innovate, grow, and learn

SOURCE: Adapted from Rajendra and Nikiforova, 2016.
RAND *RR1930-5.4*

While incentives are generally viewed as positive motivators, some organizations we interviewed used adverse means to ensure the workforce was amenable to participating in efforts that supported the learning culture. These adverse means included reprimands, removals, and/or identification of people who were not supportive of the learning culture. These people were given the opportunity to correct themselves or find different employment.

Table 5.1
IFC Lessons-Learned Criteria

Examples of KNOWbel Submission Form Using Score	
Criterion	Examples
1. Processes	Describe what methods were employed at various stages of the project, program, or initiative and how this facilitates one or more of the following: • CONNECT: Responded to a client issue/challenge by providing an integrated knowledge solution (across departments in the World Bank Group) • SHARE: Delivered value, solutions, and impact to external clients by utilizing the World Bank Group's stakeholder/industry knowledge and expertise
2. Results	a. What results have already been achieved as a result of your knowledge sharing/capture/transfer/reuse and/or collaboration efforts? b. How have you been able to keep track of this?
3. Scalability and lessons learned	a. What challenges did you face in deploying the approach you used? b. How would you deploy the project/program/initiative in the future so that it has a wider reach? c. What are some lessons learned?

SOURCE: Adapted from Rajendra and Nikiforova, 2016.

Knowledge Management

Knowledge management, as defined earlier in this report, is

> a discipline that promotes an integrated approach to identifying, capturing, evaluating, retrieving, and sharing all of an enterprise's information assets. These assets may include databases, documents, policies, procedures, and previously uncaptured expertise and experience in individual workers.[25]

Knowledge management is part of the foundation of lessons learned; without the ability to share information, lessons learned cannot be implemented. Indeed, in interviews, government and military organizations overwhelmingly acknowledged the essential nature of knowledge management for lessons learned programs; the majority of private-sector companies also acknowledged its importance.[26]

From an IT perspective, knowledge management is a knowledge-sharing platform on which archived material can be easily accessed. These IT solutions can vary in sophistication, from libraries accessible to the entire organization to collaboration sites

---

[25] Duhon, 1998. There are many definitions of knowledge management; the one used here is most applicable to the context of this study. See also Koenig, 2012; Hussein and Khan, 2015; and Meihami and Meihami, 2014.

[26] See Appendix B.

similar to Microsoft SharePoint. Bechtel executives described their company's system to us (see box).

Successful knowledge management depends not only on a willingness to share information but also on high data quality and fidelity. Databases must be maintained. Lessons-learned reporting needs to pass quality assurance standards to make sure the information can be used by others easily. Data must be added to a knowledge management system with care; just because the system might not process corporate finances does not mean that rigorous data hygiene standards do not apply.

The topic of knowledge management does come with some practical warnings, found through the literature and interviews. For example, both CSI and the U.S. Navy warned against building databases of lessons learned. CSI contends that few people looked at its database and that databases require a lot of work to maintain and operate. CSI further contends that databases were found to be more useful for specific tasks for which a single, bounded answer exists.

The lessons-learned officials representing the U.S. Navy concurred with CSI's views and experiences. The Navy representatives caution "against being overly database-centric because you cannot compel people to use it." (See box on next page.)

Knowledge Management in Information Technology

Bechtel sees knowledge management as an IT endeavor with the goal of offering the right information to the right people at the right time to guide better business decisions. The company developed networking sites that connect people of similar skill sets across the entire organization using software that accepts suggestions and questions from the workforce and categorizes the input as recommendations, problems, or potential lessons learned. These are then vetted by SMEs before being made available on the Bechtel wiki. The key to IT initiative success is user-friendly access.

SOURCE: Interview with Bechtel executives.

Knowledge Management and IT in the U.S. Navy

The knowledge management function consists of seven contractors who conduct lessons-learned activities, with two dedicated to working with JLLIS. Another five contractors are decentralized throughout the Navy to do data-entry and validation of what is loaded into the database. This is a $1 million contract per year for knowledge management.

When asked about how this specific type of structure might apply to the Department of State, the Navy representatives suggested that an undermanned organization, such as the Department of State, would have trouble launching a database. The department could train and assist the people already in the field and could then have stakeholders elsewhere review and validate the data. The representatives suggested putting paid professionals in key data hubs to get the job done—"your people do not want to get bogged down with" data entry and upkeep, noting that Navy SEALs talk to "lessons-learned guys who put [the data] into the system."

SOURCE: Telephone interview with director, Navy Lessons Learned, June 23, 2016.

CHAPTER SIX

Findings, Conclusions, and Recommendations

Chapters One through Five highlighted requirements and enablers to implement a successful lessons-learned program. Building on the preceding discussions, this chapter provides an assessment of lessons-learned strategies other organizations have adopted that the Department of State should consider as it moves forward with its enterprise solutions for lessons-learned practices. It also discusses potential obstacles to implementation and corresponding mitigation strategies.

Findings

We found that best practices differ, often greatly, depending on the strategic vision or approach an organization selects. We found that vision and approach matter far more than whether the organization is in the private or public sector. For example, both NASA and Bechtel rely on centrally organized corporate learning strategies, value the learning benefits of failure and mistakes, and maintain robust knowledge-management systems. This finding in itself should empower any organization, whether government agency or private firm, with the confidence that strong lessons-learned programs are within its reach. The following are additional findings:

1. The designation of an integrating or steering element (office, department, other group) helps organizations drive lessons-learned efforts.
2. Organizations can use three primary strategic approaches to drive lessons-learned programs and/or the transfer of knowledge:
   a. BPM (structured learning executed at all levels of an organization, led by managers, as part of their day-to-day responsibilities)
   b. corporate learning strategies (a separate institution within an organization that helps the entire organization acquire, apply, and share knowledge[1])
   c. a hybrid of both.

---

[1] International Institute for Management Development, 2011.

4. Developing a vision for the lessons-learned program is key to its success. The vision is necessary to answer essential implementation decisions:
    a. What are the desired outcomes of the program?
    b. What approaches will be used to achieve these outcomes (BPM and/or corporate learning strategies)?
    c. Whom do you wish to reach in the organization (individuals through leadership and/or business units)?
    d. How will the lessons-learned program be governed (centralized or decentralized) based on desired outcomes and target audience?
5. Many approaches to collection, validation, and dissemination exist. Deliberate choices need to be made on these matters and in general to clearly guide the growth of lessons learned.
6. A strong culture of learning is essential for a lessons-learned program to function effectively and can be fostered through intraorganizational enabling elements, such as leader and workforce buy-in; establishment of practices, policies, and procedures; and the creation and/or connection of existing organizational infrastructure.
7. *Extra*organizational knowledge sources, including communities of practice and informal networking, can enhance the effectiveness of the lessons-learned program.
8. Many challenges to creating the culture of learning exist, including reticence to share failures, resource constraints (personnel and/or funding), and the need to demonstrate value added or impact.
9. *Intra*organizational elements, coupled with open and inclusive collection methods, can help leaders hedge against undue bias or influence.
10. Knowledge management—making collected information and lessons that have been learned available to the broader organization—is a major aspect of the lessons-learned effort.

Reflections on the Department of State's Current Lessons-Learned Enterprise

As previously mentioned, we did not aim to assess the Department of State's current lessons-learned capabilities or seek to prescribe a tactical implementation plan. Instead, this report offers relevant thematic best practices for the department to consider when developing its next steps. To ensure that our recommendations fit the context of Department of State operations, we familiarized ourselves with the department and many of its lessons-learned components as part of our data collection.

In addition to CSCD's learning activities, several Department of State centers and offices conduct lessons learned or are involved in activities similar to those that are essential to a robust learning culture. Examples include Diplomatic Security, which works to ensure the safety of diplomats; the office of the Transparency Coordinator, which is improving document preservation and transparency systems, a key component of knowledge management; and the Operations Center's Office of Crisis Management and Strategy, which, among other things, coordinates the department's responses to crises.

The small sample of organizations we interacted with or interviewed helped inform our understanding of Department of State's current environment, but we did not conduct a comprehensive assessment. We recommend that the department consider conducting an exhaustive assessment of its own current lessons-learned capabilities as a preliminary step for developing its enterprise lessons-learned program. Such an assessment should include all current lessons-learned activities to extract best practices already in use. Consideration of the existing lessons-learned structure would promote buy-in and allow the department to build on what is already in place.

The Maturity Implementation Model

The existing mixture of decentralized and formal, centralized lessons-learned artifacts in the Department of State can be best represented by the Maturity Implementation Model, a tool one of our private-sector interviewees created and shared that measures and describes an organization's status in its journey toward implementing a lessons-learned program (see Table 6.1). The model suggests that the department is firmly in the third of five levels of implementing a culture of organizational learning. We based this assessment on the contextual knowledge and understanding we developed over the course of the research. We did not interview all groups within the Department of State involved in lessons-learned activities, but our aim in including this model here is to offer the department a possible model for future assessments.

Table 6.1
Department of State Implementation of Culture of Organizational Learning

Level	Description
5. Continuous improvement	Learning organization committed to improvement
4. Performing	Demonstrated consistent performance
3. Established	Confirmed organizational commitment
2. Managing	Implementing best practices and demonstrating value
1. Emerging	Demonstrated management commitment

SOURCE: Interview with lessons-learned coordinator at a global oil company, May 4, 2016.

The demonstrated management commitment indicative of a level 1 organization is prevalent in the Department of State; this research effort is itself an example of that commitment. The methodological collection rigor required to produce policy implementation reviews (PIRs) demonstrates a start at using best practices in the formal lessons-learned program, necessary for a level 2 ranking. Furthermore, the incorporation of PIR lessons into the FSI curriculum reflects the impact and organizational value added. The historic attempts to initiate lessons-learned activities and the current decentralized, formal approaches found across the department, combined with the centralized, formal capabilities FSI and other elements offer, place the Department of State firmly in level 3.

To reach level 4, the Department of State needs to continue expanding its lessons-learned enterprise and activities. To achieve level 5, full learning organization status, the department as a whole will need to have adopted and embraced a learning culture and a mindset of continuous improvement. The actionable recommendations offered in the next section are prescriptive in advancing the department's implementation efforts.

Advancing Lessons-Learned Implementation in the Department of State

The Department of State has taken major strides to grow its enterprisewide lessons-learned capability, and this subsection provides a more granular planning framework for next steps the department should consider. This framework, which we derived from our research, the literature, and expert opinion, guides planners to consider formalization, governance, human capital management, and resources. This framework is unique in that it allows for a holistic capabilities assessment across the multiple functional domains. Table 6.2 summarizes focal areas we based on policy assessments and field research for advancing implementation of the department's lessons-learned program. In column three, checks indicate activities the department is already implementing, and Xs indicate areas on which it should focus. It is important to note that this is not a comprehensive analysis of all the Department of State's lessons-learned capabilities—our intent is to provide a framework that the department and other organizations can use to guide planning.

Formalization

The QDDR provides the Department of State with guiding fundamental principles that are indeed strategic. The QDDR explicitly states Secretary of State–directed imperatives—such as creative problem-solving and institutionalizing policy to encourage innovation while managing risk—and anchors them to the capturing and communicating of lessons learned. As with doctrine in any organization, however, the department does not provide the details for strategic and tactical implementation and subsequent execution and metering. These aspects of program implementation must also be developed. As of this writing, the department is developing the specific implementation strategies to build up to a steady-state level of execution. CSCD is on a

Table 6.2
Lessons-Learned Capability Assessment Framework

Focal Areas	Organizational Attributes	Department of State Practices	Next Steps
Formalization	Written documentation, e.g., policies and procedures	✓ The QDDR provides visionary framing for intent of the lessons-learned initiatives. X Policy does not exist to promote or enforce audience-based lessons-learned activities. X Policy not exist to drive actor activities.	• Develop a vision for the specific implementation plan to set conditions for program execution. • Develop policies that promote and enforce the establishment of a learning culture. • Develop actor policies to standardize and institutionalize the capability.
Governance	Designating roles and responsibilities and leadership involvement	✓ The intraorganizational units are in place to enable an enhanced lessons-learned program. X These units, however, are not yet all aligned for optimal lessons-learned processing. X A single lead structure with primary responsibility has not been identified. ✓ Departmental leadership is invested and demonstrates the willpower to succeed.	• Identity a lead organization or body with primary responsibility for enterprisewide lessons learned—perhaps even a CoE approach. • In accordance with the vision, identify or create the organizational units needed to drive the lessons-learned program. • Leadership should continue to establish a vision for the program and create infrastructure and mechanisms where needed and/or connect elements that already exist.
Human capital management	Training, talent management, and culture	X The lessons-learned initiatives are not yet mature enough for formal training of actor or audience. ✓ CSCD staff is using employee-identified best practices as actors to produce PIRs. ✓ Dedicated staff for key lessons learned activities exists (CSCD, transparency coordinator, archivist, etc.). X Cultural barriers to workforce buy-in exist.	• As part of the marketing campaign, develop techniques for audience training. • Standardize training for actors for such activities as data collection, validation, and preparation of deliverables. • In accordance with vision, determine actor requirements and source (U.S. government, contractor, etc.) • Create motivational and mechanism-based strategies to drive culture of learning.
Resources	Funding and facilities	✓ Funding is seemingly available to meet vision. ✓ FSI has an existing facility to house lessons learned CoE-like presence and/or to conduct lessons-learned training or activities.	• Develop a budget to inform the near and long-term vision for the lessons-learned program. • Determine the space required for lessons-learned functions and use existing infrastructure.

NOTE: CoE = center of excellence.

journey to drive cultural changes, while also providing tools to enable the diplomatic mission.

Policy is defined as "a high-level overall plan embracing the general goals and acceptable procedures especially of a governmental body."[2] A key point with policy is that it is authoritative and serves as a formal, institutionalized mechanism to drive and enforce compliance. The Department of State should call on all managers at every level to help develop and bolster the policy platform that will drive the lessons-learned program.

Policies can direct both actors and audience in the lessons-learned efforts. As previously mentioned, policies and procedures should also be used to drive the culture of learning the Department of State seeks. This is particularly the case where the lessons-learned and knowledge-sharing missions converge.

Governance

The Department of State has several formal intraorganizational elements in place to execute the lessons-learned mission. What the department lacks, however, is a vision for a plan of action to guide the implementation of the program. This plan of action should be predicated on the overall vision of what the program is to achieve. Of equal importance for organizational constructs is that the department does not have a dedicated cell with decision authority to integrate all the lessons-learned activities and capabilities.

The Department of State also needs to establish a governing or steering body to make lessons-learned and/or knowledge-sharing decisions. The caution here is not to establish boards or work groups that do not have the ability to authoritatively reach or dictate to the entire department. An organization with primary responsibility will have the ability to span the department and act as the center of gravity for all centralized lessons-learned matters. The U.S. military sometimes uses the *center of excellence (CoE)* concept for such activities. Academia has leveraged the U.S. Army definitions to derive a generalized definition and critical dimensions of such a center. Fisher & Garcia-Miller define the CoE as follows:

> [a] premier organization providing an exceptional product or service in an assigned sphere of expertise and within a specific field of technology, business, or government, consistent with the unique requirements and capabilities of the CoE organization.[3]

---

[2] "Policy," in *Merriam-Webster Dictionary*, online ed., undated.

[3] TRADOC definition, as quoted in William Craig, Matt Fisher, Suzanne Garcia-Miller, Clay Kaylor, John Porter, and Scott Reed, "Generalized Criteria and Evaluation Method for Center of Excellence: A Preliminary Report," technical note, Pittsburgh, Pa.: Carnegie Mellon University, Software Engineering Institute, December 2009.

Human Capital Management

With the establishment of CSCD, the Department of State has appointed a cadre of actors dedicated to the lessons-learned function. Once the vision for the reach of the lessons-learned effort is set in place, the department can then determine the personnel requirements it will take to meet these demands. For example, if the department wants a more decentralized formal approach, appropriate personnel need to be in place in disparate locations to achieve this goal, similar to FEMA's and NASA's lessons-learned programs. A plan of action determines the personnel requirement and should take resources into account to ensure the plan is executable.

It is noteworthy to mention that staffing requirements can be outsourced or insourced for an organization once the personnel requirements and associated budgets are determined. The outsourced options typically involve contractors; other options include borrowing personnel from another organization and using interns. CCO uses temporary unpaid interns to write and publish *PRISM* magazine, which reports on CCO activities and research.[4] Insourcing options include establishing permanent or temporary positions or requirements. These positions can then be filled with new hires, borrowed internal personnel, or people temporarily available due to rotation.

As for a personnel-related qualitative assessment, the Department of State workforce's culture is again noteworthy; the department itself has expressed significant concerns related to adopting a learning culture. Both marketing and training to increase willingness to embrace change in general and sharing knowledge can mitigate the barriers associated with culture. Policies and procedures can help encourage knowledge sharing. Some organizations include criteria to bolster the culture of innovating and learning in job and duty descriptions, performance standards and objectives, and performance reviews. Incentives are another means of encouraging employees to participate in activities that promote a culture of learning and knowledge sharing.

Two levels of training exist from an organizational perspective. The first is training for the trainers or, in this case, the actors in the lessons-learned process. CSCD is investigating ways to enhance its deliverables in terms of data collection and product dissemination. The second type of organizational training is for the audience or recipients of the lessons-learned deliverables. Department of State interviewees cited several barriers to creating a learning culture, including a need-to-know mentality with respect to knowledge sharing, a high frequency of rotation across all levels of the organization, and an individualist mentality within the department. Training is one form of institutional offering that can help motivate cultural change.

Resources

We define *resources* as consisting of funding and facilities. Few, if any, organizations have unlimited funding. Leaders are forced to prioritize. The funding decisions allocate

---

[4]   Interview with CCO, Washington, D.C., August 19, 2016.

money to pay for the programs and any associated personnel, equipment, real property, etc. The allocation of funding in any organization typifies what the organization deems important. A true commitment to a cause is usually embodied by dedicated staff with an authorized funding line. Program spending also includes training, as discussed earlier.

Programs clearly need to market themselves to be competitive in the intraorganizational prioritization of funding, although marketing activities are not free. A good marketing campaign for lessons-learned programs should demonstrate the value these types of offerings provide to the organization. While measurement is sometimes difficult in business units responsible for less quantifiable outputs, it is still necessary in terms of defending or justifying the allocation of funding. Again, measuring and monitoring program activities also produce costs.

The Department of State has a dedicated facility for FSI that can serve as the center of gravity for the lessons-learned program. The mission of the facility matches the training roles suited for both actors and audience. Additionally, trainees or students and/or their family members are available to serve in the lessons-learned mission as they rotate through for training. Family members sometimes have skill sets that can be leveraged, thereby increasing the available labor pool in the lessons-learned mission.

Thematic Best Practices

We offer the following thematic best practices to guide the Department of State as it continues its lessons-learned journey:

1. *Develop an annual collection plan that flows from the organization's vision and mission, and get senior leadership buy-in on what is to be collected and disseminated.* This approach helps determine the offerings of the lessons-learned capability and sets conditions for staffing and resourcing decisions for the lessons-learned program. Collection planning should include determining triggers or learnable events and when to collect on these events. Collection could occur at all stages of an employee's assignments or career progression, for example, at the end of FSI training, mid–embassy tour or ad hoc, and on returning from an assignment. This practice would allow department employees to be consumers during training and consumers or producers once in the field.

2. *Determine the specific outcomes lessons-learned initiatives should achieve to establish a tangible list of requirements.* Chapter Two provided several examples of different categories of goals, including institutional knowledge and risk avoidance. Once the requirements to achieve the desired outcomes are established, identify the current centralized and decentralized and formal and informal capabilities currently in place to determine where additional capability needs to be created

and also where existing capability can be integrated into an enterprisewide program.

3. *To better govern the lessons-learned program, assign an office or body primary responsibility.* Clear leadership in this sense can help the organization remain focused on the program's vision and goals. Our research has shown that organizations use formalized products, such as policies, performance reviews, and incentive programs, to drive the culture of learning. The Department of State should consider implementing a combination of these formal tools to shape its learning culture.

4. *Establish a formal, named campaign with accompanying slogan to promote the culture of learning and the existence of the formalized lessons-learned capability.* Interviewed organizations—such as the CIA's CSI—developed robust marketing campaigns that included posters physically hung around their buildings.

5. *Develop a resource-allocation strategy to ensure resources are not a constraint to the lessons-learned enterprise's ability to meet the desired outcomes of the program.* Begin with personnel analysis to determine the number and type of staff required and then use innovative ways to insource and outsource those requirements.

Interview Protocol, Research Lenses, and Structured Discussion Protocol

The semistructured interviews included the questions below. Over the course of the study, the research lenses, also below, emerged as key themes. We used these research lenses to guide the roundtable discussions. The structured discussion protocol reorganizes the interview questions according to the research lenses.

Interview Questions—Version 1 (March 29, 2016)

Lessons Learned Best Practices Study
Why Do Lessons Learned

- Why do you conduct lessons learned reviews? What benefits do these reviews generate in terms of impact, efficiency, personnel development, etc.?
- Are lessons learned reviews intended to generate near-term or long-term benefit? Are they expected to yield program-specific improvements or broader generalized improvements?

Conduct—Management of Lessons Learned Reviews

- Are certain preconditions necessary to learn lessons from past performance effectively?
  - Examples might include leadership support, safety from retaliation, structured learning or training processes
- How do you decide what to review?
- Who decides what is worthy of a review?
- Do decisions come from leadership or the working level, or from elsewhere?
- Who leads a review, and who participates?
- Are lessons-learned reviews centrally managed and conducted, or can business units undertake their own reviews?

    – If the process is decentralized, how do you ensure that all reviews follow consistent procedures, standards, formats, etc.?

Conduct—Execution of Lessons-Learned Reviews

- How do you ensure that lessons-learned reviews are comprehensive, fact based, and free from undue bias or influence?
- How do you gather information and insights (e.g., document review, interviews, surveys, roundtable discussions)?
- What ground rules govern lessons-learned reviews?
- How do you ensure people are willing to discuss failures?
- How do you assess functions in which responsibilities are divided among multiple components of the organization?

Dissemination of Lessons Learned

- How are lessons-learned assessments disseminated?
- Who is in charge of dissemination?
- How important is a versatile IT infrastructure for use and dissemination of lessons learned?
- Are lessons incorporated into professional training? If so, how and toward what ends?
  - Examples might include onboarding training, periodic training for certain jobs or levels, informal discussions
  - Applications might be to train managers, promote team building, promote critical thinking, enhance project management

Implementation

- How do organizations use the results of lessons-learned studies to change practices and behavior?
- Does someone specific own the lessons and their implementation?
- Does leadership specifically endorse lessons, or is their imprimatur on the overall lessons-learned process sufficient?
- How do lessons learned lead to changes in business practices?
- What are obstacles to implementing lessons learned effectively, and how does your organization overcome them?
  - Examples might include cultural barriers, intolerance of risk or error, poor dissemination, absence of incentives to change
- How does your organization use lessons-learned reviews to foster a culture of organizational learning?

Impact

- Do lessons-learned assessments specify metrics to assess the impact of implementing lessons?
- Are lessons revisited after a certain amount of time to see if they were implemented and if they had the desired impact? Is this process the responsibility of the lessons-learned team or the affected business unit?

Research Lenses

1. *Vision* covers establishing a vision for the lessons-learned program; includes expected outcomes and target audience
2. *Enabling elements* covers necessary elements to enable a lessons-learned program; includes culture of learning, leader buy-in, organizational structure, etc.
3. *Collection planning* covers what you collect and how that it determined
4. *Collection methodology* covers how you collect and what do you do with the data; includes methodological approaches for data collection, validation, and deliverables
5. *Impact* describes forms of impact the lessons-learned programs have had and how they achieve or maintain credibility; includes informing decisions, policy and guidance changes, curriculum for training and education, new procurements, populating a knowledge-sharing system, etc.
6. *Knowledge management* covers approaches to knowledge management; includes staffing, training lessons-learned program participants, etc.
7. *Talent management* covers staffing (dedicated and matrix) and training
8. *Communities of practice* covers the importance of networking across the community—both internally and externally to the organization
9. *Overcoming obstacles* covers overcoming barriers to the program; includes concerns about professional well-being in the wake of adversity, organizational positioning, funding, etc.

Structured Discussion Protocol

Why Do Lessons Learned

- Why do you conduct lessons learned reviews? What benefits do these reviews generate in terms of impact, efficiency, personnel development, etc.?
- Are lessons learned reviews intended to generate near-term or long-term benefit? Are they expected to yield program-specific improvements or broader generalized improvements?

Vision

Establishing a vision for the lessons learned program; expected outcomes and target audience

- How do you decide what to review?
- Who decides what is worthy of a review?
  - Do decisions come from leadership or the working level, or from elsewhere?
- Who leads a review, and who participates?
- Are lessons learned reviews centrally managed and conducted, or can business units undertake their own reviews?
  - If the process is decentralized, how do you ensure that all reviews follow consistent procedures, standards, formats, etc.?

Enabling Elements

Necessary elements to enable a lessons-learned program; culture of learning, leader buy-in, etc.

- Are certain preconditions necessary to learn lessons from past performance effectively?
  - Examples might include leadership support, safety from retaliation, structured learning/training processes, etc.
- Does leadership specifically endorse lessons, or is their imprimatur on the overall lessons-learned process sufficient?
- How do you ensure people are willing to discuss failures?

Collection Planning

On what do you collect and how is it determined; collection plans

- How do you assess functions in which responsibilities are divided among multiple components of the organization?
- What ground rules govern lessons-learned reviews?
- How do you ensure that lessons-learned reviews are comprehensive, fact based, and free from undue bias or influence?
- What organizational aspects do you have in place (corporate learning and/or management tradition)?

Approaches to Collection, Validation, and Dissemination

How do you collect and what do you do with that data; methodological approaches for data collection, validation, and deliverables

- How do you gather information and insights (e.g., document review, interviews, surveys, roundtable discussions)?

- Describe your validation strategies
- How are lessons-learned assessments disseminated?
- Who is in charge of dissemination?
- Do you use utilize push or pull strategies?
  - Ex: efficiency competitions
  - Suggestion boxes

Impact

Describe forms of impact the lessons-learned programs have had and how they achieve or maintain credibility; inform decisions, policy and guidance changes, curriculum for training and education, and new procurements; populate a knowledge-sharing system; etc.

- How do organizations use the results of lessons-learned studies to change practices and behavior?
- How do lessons learned lead to changes in business practices?
- Does someone specific own the lessons and their implementation?
- Do lessons-learned assessments specify metrics to assess the impact of implementing lessons?
- Are lessons revisited after a certain amount of time to see if they were implemented and if they had the desired impact? Is this process the responsibility of the lessons-learned team or the affected business unit?
- How do you measure impact?
- Are lessons incorporated into professional training? If so, how and toward what ends?
  - Examples might include onboarding training, periodic training for certain jobs or levels, informal discussions
  - Applications might be to train managers, promote team building, promote critical thinking, enhance project management

Knowledge Management

Discuss approaches to knowledge management; staffing, training lessons-learned program participants, etc.

- How do you intend to inculcate a learning culture across levels of the organization (e.g., individual, section or team, leadership, and/or dedicated staff)?
- Do you intend to track participation in the program (e.g., via online hits by grade/rank)?
- How important is a versatile IT infrastructure for use and dissemination of lessons learned?

Talent Management

Staffing (dedicated and matrix) and training

- Entire work force vs. dedicated "lessons-learned staff" vs. management or hybrid model?
- Describe human capital sourcing solutions for lessons-learned function.

Communities of Practice

Discuss the importance of networking across the community.

Overcoming Obstacles

Overcoming barriers to the program; concerns about professional well-being in the wake of adversity, organizational positioning, funding, etc.

- What are obstacles to implementing lessons learned effectively, and how does your organization overcome them?
  - Examples might include cultural barriers, intolerance of risk or error, poor dissemination, absence of incentives to change
  - Resource constraints
  - Organizational positioning

Interview Findings

Interviewee data, often anonymous, appear throughout this report in the forms of direct quotes, vignettes, and/or case studies to support key concepts. Table B.1 details the frequency with which interviewees suggested particular factors contributed to or occurred in the context of their lessons-learned programs. Frequency is tabulated according to sector—military organizations, government and public sector, and private sectors, but specific contributors are not identified.

Table B.1
Interview Findings

Factor	Military (*n* = 3)	Government (*n* = 6)	Private Sector (*n* = 16)	Total (*n* = 35)
Vision	13	6	7	26
Learning culture	13	5	14	32
Leader buy-in	13	5	12	30
Collection planning and predetermined triggers for lessons-learned reviews	8	2	7	17
Knowledge management	13	6	8	27
IT-based database	11	5	4	20
Organization-specific lessons-learned portal	8	4	5	17
Shared or interagency lessons-learned portal	10	2	0	12
Lessons-learned portal (all types)	11	5	5	21
Human capital planning	13	5	10	28
Communities of practice	11	4	6	21
Overcoming obstacles	11	2	2	15
Resourcing and staffing challenges	7	1	3	11

In categorizing the factors, we asked the following questions about interview responses:

1. Vision: Does the organization or interviewee link lessons-learned programs to organizational vision or mission?
2. Learning culture: Does the organization/interviewee explicitly state the value of a learning culture?
3. Leader buy-in: Does the organization or interviewee explicitly state the value of leader buy-in?
4. Collection planning and predetermined triggers for lessons-learned reviews: Does the organization have predetermined triggers for a lessons-learned review process (e.g., annual collection planning, specific triggers)?
5. Knowledge management: Does the organization incorporate knowledge management into its lessons-learned review processes and protocols?
6. IT-based database: Does the organization utilize an IT database as a part of its lessons-learned collection and/or dissemination processes?
7. Organization-specific lessons-learned portal: Does the organization have an internal lessons-learned portal?
8. Shared or interagency portal: Does the organization have a shared, interorganizational lessons-learned portal (e.g., JLLIS)?
9. Lessons-learned portal (all types): Does the organization have any type of lessons-learned portal?
10. Human capital management: Does the organization or interviewee note the importance of talent management (including training) for lessons-learned processes?
11. Communities of practice: Do the organization's lessons-learned practitioners engage in lessons-learned communities of practice?
12. Overcoming obstacles: Were interviewees open about the challenges or obstacles of undertaking successful lessons-learned programs?
13. Resourcing and staffing challenges: Does the organization report a need for additional lessons-learned staff?

One limitation of our approach is that, to ensure broader applicability of the findings, we drew our conclusions from small samples of organizations from various fields; however, these sample organizations may not necessarily be representative of the organizational fields in which they exist. Larger samples should be considered for a more comprehensive understanding of the trends in lessons-learned initiatives inside a specific guild.

Abbreviations

AAR	after-action review
BPM	business process management
BPR	business process reengineering
CALL	Center for Army Lessons Learned
CCO	Center for Complex Operations
CIA	Central Intelligence Agency
CIO	chief information officer
CLO	chief learning officer
CoE	center of excellence
CSCD	Center for the Study of the Conduct of Diplomacy
CSI	Center for the Study of Intelligence
DoD	Department of Defense
FEMA	Federal Emergency Management Agency
FSI	Foreign Service Institute
GAO	Government Accountability Office
GE	General Electric
JLLIS	Joint Lessons Learned Information System
IFC	International Finance Corporation
IT	information technology
JPL	Jet Propulsion Laboratory
NASA	National Aeronautics and Space Administration
NATO	North Atlantic Treaty Organization
PIR	policy implementation review
QDDR	Quadrennial Diplomacy and Development Review
SME	subject-matter expert
UPS	United Parcel Service
USAID	U.S. Agency for International Development
USAF	U.S. Air Force
USMC	U.S. Marine Corps

References

Adler, Paul S., Barbara Goldoftas, and David I. Levine, "Flexibility Versus Efficiency? A Case Study of Model Changeovers in the Toyota Production System," *Organization Science*, Vol. 10, No. 1, February 1999, pp. 43–68. As of February 8, 2017:
http://pubsonline.informs.org/doi/abs/10.1287/orsc.10.1.43

Argote, Linda, and Paul Ingram, "Knowledge Transfer: A Basis for Competitive Advantage in Firms," *Organizational Behavior and Human Decision Processes*, Vol. 82, No. 1, 2000, pp. 150–169. As of February 8, 2017:
http://dx.doi.org/10.1006/obhd.2000.2893

Argyris, Chris, *On Organizational Learning,* 2nd ed., Oxford, U.K.: Blackwell Business, 1999.

Avis, Peter, and Joe Sharpe, "Operationalization of the Lessons Learned Process: A Practical Approach," in Susan McIntyre, Kimiz Dalkir, and Irene C. Kitimbo, eds., *Utilizing Evidence-Based Lessons Learned for Enhanced Organizational Innovation and Change*, Washington, D.C.: IGI Global, 2015.

Ben-Hur, Shlomo, Bernard Jaworski, and David Gray, "Aligning Corporate Learning with Strategy," *MIT Sloan Management Review*, September 1, 2015. As of February 8, 2017:
http://sloanreview.mit.edu/article/aligning-corporate-learning-with-strategy/4/

Britton, Bruce, "Organizational Learning in NGOs: Creating the Motive, Means, and Opportunity," Oxford, U.K.: International NGO Training and Research Centre (INTRAC), Praxis Paper No. 3, March 2015. As of June 8, 2016:
http://www.intrac.org/data/files/resources/398/
and Michael Rosemann, eds., *Handbook on Business Process Management*, 2nd ed., Berlin: Springer-Verlag, 2015.

CALL—*See* Center for Army Lessons Learned.

Center for Army Lessons Learned, *Establishing a Lessons Learned Program: Observations, Insights, and Lessons*, Fort Leavenworth, Kan., Handbook 11-33, June 2011. As of February 8, 2017:
http://usacac.army.mil/sites/default/files/publications/11-33.pdf

Center for Army Lessons Learned, *CALL Services*, Fort Leavenworth, Kan., Handbook 15-11, June 2015. As of February 8, 2017:
http://usacac.army.mil/sites/default/files/publications/15-11.pdf

Center for Leadership Studies, "Situational Leadership," web page, undated. As of February 8, 2017:
https://situational.com/the-cls-difference/situational-leadership-what-we-do/

Centers for Disease Control and Prevention, "CDC Unified Process Practices Guide: Lessons Learned," Atlanta, Ga., November 30, 2006. As of February 8, 2017:
https://www2.cdc.gov/cdcup/library/practices_guides/CDC_UP_Lessons_Learned_Practices_Guide.pdf

Commandant Publication P3120.17B, *U.S. Coast Guard Incident Management Handbook*, May 2014. As of April 17, 2017:
https://www.uscg.mil/d9/D9Response/
docs/USCG%20IMH%202014%20COMDTPUB%20P3120.17B.pdf

Craig, William, Matt Fisher, Suzanne Garcia-Miller, Clay Kaylor, John Porter, and Scott Reed, "Generalized Criteria and Evaluation Method for Center of Excellence: A Preliminary Report," technical note, Pittsburgh, Pa.: Carnegie Mellon University, Software Engineering Institute, December 2009. As of April 18, 2018:
http://resources.sei.cmu.edu/library/asset-view.cfm?assetid=8971

Daft, Richard L., *Organization Theory and Design*, 3rd ed., St. Paul, Minn.: West Publishing Company, 1989.

Duhon, Bryant, "It's All in Our Heads," *Inform*, Vol. 12, No. 8, September 1998, pp. 8–13.

Ernst and Young, "Profiting from Experience: Realizing Tangible Business Value from Programme Investment with Lessons Learned Reviews," pamphlet, London, 2007. As of April 26, 2017:
http://www.ncc.co.uk/download.php?647162424a63365671664b395948303163743256661784e71664
b654c4a3176656b667377677a464d5a79324968513d3d

Fiumara, Karen, "Safety Reporting Leads to Safer Systems," Safety Matters blog, Brigham and Women's Hospital, February/March 2016. As of May 8, 2017:
https://bwhsafetymatters.org/safety-reporting-leads-to-safer-systems/

Frost, Alan, "Introducing Organizational Knowledge," KMT: Knowledge Management Tools website, 2010. As of May 8, 2017:
http://www.knowledge-management-tools.net/introducing-organizational-knowledge.html

Garvin, David, Amy C. Edmondson, and Francesca Gino, "Tool Kit: Is Yours a Learning Organization?" *Harvard Business Review*, March 2008, pp. 2–3.

Grabowski, Martha, and Karlene Roberts, "Risk Mitigation in Large-Scale Systems: Lessons from High Reliability Organizations," *California Management Review*, Vol. 39, No. 4, 1997, pp. 152–162. As of February 8, 2017:
http://cmr.ucpress.edu/content/39/4/152.abstract

Hammer, Paul, "What Is Business Process Management?" in Brocke and Rosemann, 2015.

Harmon, Paul, "The Scope and Evolution of Business Process Management," in Brocke and Rosemann, 2015.

Harrington, H. James, *Business Process Improvement: The Breakthrough Strategy for Total Quality, Productivity, and Competitiveness*, Boston: McGraw-Hill, 1991.

Huang, Jimmy C., and Sue Newell, "Knowledge Integration Processes and Dynamics Within the Context of Cross-Functional Projects," *International Journal of Project Management*, Vol. 21, No. 3, April 2003, pp. 167–176. As of February 8, 2017:
http://www.sciencedirect.com/science/article/pii/S0263786302000911

Hussein, Tanvir, and Salim Khan, "Knowledge Management an Instrument for Implementation in Retail Marketing," *MATRIX Academic International Online Journal of Engineering and Technology*, Vol. 3, No. 1, April 2015. As of February 8, 2017:
http://maioj.org/data/documents/april2015/041501.pdf

International Institute for Management Development, "Corporate Learning," Insights@IMD, No. 4, Lausanne, Switzerland, May 2011. As of February 8, 2017:
http://www.imd.org/research/publications/upload/4-Corporate-Learning-discovery-event-w-no-05-11-12-2.pdf

Investopedia, website, undated. As of April 20, 2017:
http://www.investopedia.com/

Jeon, Jongwoo, "Success Factors for a Lessons-Learned System in a Construction Organization," *Cost Engineering*, Vol. 51, No. 5, May 2009.

Koenig, Michael E. D., "What Is KM? Knowledge Management Explained," *KM World*, May 4, 2012. As of February 8, 2017:
http://www.kmworld.com/Articles/Editorial/What-Is-.../What-is-KM-Knowledge-Management-Explained-82405.aspx

Leitner, Karl-Heinz, and Campbell Warden, "Managing and Reporting Knowledge-Based Resources and Processes in Research Organisations: Specifics, Lessons Learned and Perspectives," *Management Accounting Research*, Vol. 15, No. 1, March 2004, pp. 33–51.

Lunenburg, Fred C., "Organizational Structure: Mintzberg's Framework," *International Journal of Scholarly, Academic, Intellectual Diversity*, Vol. 14, No. 1, 2012. As of February 8, 2017:
https://platform.europeanmoocs.eu/users/8/
Lunenburg-Fred-C.-Organizational-Structure-Mintzberg-Framework-IJSAID-V14-N1-2012.pdf

Marlin, Mark, "Implementing an Effective Lessons Learned Process in a Global Project Environment," *Annual Project Management Symposium Proceedings*, Dallas, Tex., 2008. As of February 8, 2017:
http://www.westney.com/wp-content/uploads/2014/05/
Implementing-an-Effective-Lessons-Learned-Process-In-A-Global-Project-Environment.pdf

McIntyre, Susan, *Utilizing Evidence-Based Lessons Learned for Enhanced Organizational Innovation and Change*, Hershey, Pa.: IGI Global, 2015.

Meihami, Bahram, and Hussein Meihami, "Knowledge Management a Way to Gain a Competitive Advantage in Firms (Evidence of Manufacturing Companies)," *International Letters of Social and Humanistic Sciences*, Vol. 14, 2014, pp. 80–91. As of February 8, 2017:
https://www.ceeol.com/search/article-detail?id=168572

Mercy Corps, "Our Mission," web page, undated. As of May 18, 2016:
https://www.mercycorps.org/about-us/our-mission

Mercy Corps, "Managing Complexity: Adaptive Management at Mercy Corps," Portland, Oreg., January 2012a. As of June 1, 2016:
https://www.mercycorps.org/sites/default/files/Adaptive%20management%20paper_external.pdf

Mercy Corps, *Program Management Manual,* Portland, Oreg., January 2012b. As of June 1, 2016:
https://www.mercycorps.org/sites/default/files/programmanagementmanualpmm.pdf

Mercy Corps, "Practicing What We Preach: A Review of Learning and Research Utilization," brief, Portland, Oreg., May 2015. As of June 1, 2016:
https://d2zyf8ayvg1369.cloudfront.net/sites/default/files/
MC%20Research%20Utilization%20Information%20Summary%20%282%29.pdf

Merriam, Sharan B., and Phyllis M. Cunningham, eds., *Handbook of Adult and Continuing Education*, San Francisco: Jossey-Bass Publishers, 1989.

Milton, Nick, *Lessons Learned Handbook: Practical Approaches to Learning from Experience*, Oxford, UK: Chandos Publishing, 2010.

Murray, Art, and Jeff Lesher, "The Future of the Future: Breaking the Lessons-Learned Barrier," *KM World*, August 31, 2008. As of February 8, 2017:
http://www.kmworld.com/Articles/Column/The-Future-of-the-Future/The-Future-of-the-Future-Breaking-the-lessons-learned-barrier-37336.aspx

NASA—*See* National Aeronautics and Space Administration.

National Aeronautics and Space Administration, Office of the Chief Knowledge Officer, "Sharing Masters with Masters," undated. As of February 8, 2017: http://km.nasa.gov/knowledge-sharing/masters-with-masters/

NATO—*See* North Atlantic Treaty Organization.

North Atlantic Treaty Organization, *Allied Joint Doctrine for the Conduct of Operations*, AJP-3(B), Brussels, March 2011. As of February 8, 2017: http://nso.nato.int/nso/zPublic/ap/ajp-3(b).pdf

North Atlantic Treaty Organization, Joint Analysis and Lessons Learned Centre, "NATO Lessons Learned Process," web page, undated. As of May 8, 2017: http://www.jallc.nato.int/activities/nllprocess.asp

Oberhettinger, David, "Lessons Learned from Soup to Nuts," briefing, Pasadena, Calif.: Office of the Chief Engineer, Jet Propulsion Laboratory, California Institute of Technology, August 23, 2011.

Panetta, Leon E., and Jeremy Bash, "The Former Head of the CIA on Managing the Hunt for Bin Laden," *Harvard Business Review*, May 2, 2016. As of April 18, 2017: https://hbr.org/2016/05/leadership-lessons-from-the-bin-laden-manhunt

The Partnering Initiative, *The Case Study Toolbook: Partnership Case Studies as Tools for Change*, Cambridge, U.K., August 2014. As of June 8, 2016: http://thepartneringinitiative.org/wp-content/uploads/2014/08/Case_Study_Toolbook.pdf

Pedler, Mike, Tom Boydell, and John Burgoyne, "Towards the Learning Company," *Management Education and Development*, Vol. 20, No. 1, 1989, pp. 1–8.

Pittampalli, Al, "The Best Leaders Allow Themselves to Be Persuaded," *Harvard Business Review*, March 3, 2016. As of February 8, 2017: https://hbr.org/2016/03/the-best-leaders-allow-themselves-to-be-persuaded

"Policy," in *Merriam-Webster Dictionary*, online ed., undated. As of February 8, 2017: http://www.merriam-webster.com/dictionary/policy

Power, Brad, "How GE Applies Lean Startup Practices," *Harvard Business Review*, April 23, 2014.

Punke, Heather, "Why Brigham and Women's Hospital Put Medical Errors in Blog Form," *Infection Control & Clinical Quality*, February 22, 2016. As of February 8, 2017: http://www.beckershospitalreview.com/quality/why-brigham-and-women-s-hospital-put-medical-errors-in-blog-form.html

Rajendra, Sumithra, and Nadezda Nikiforova, "WBG Knowbel Awards: Building a Knowledge Culture," briefing, April 26, 2016.

Schein, Edgar H., *Organizational Culture and Leadership*, 4th ed., San Francisco, Calif.: Jossey-Bass, 2010.

Schulz, Kathryn, *Being Wrong: Adventures in the Margin of Error*, New York: Ecco, 2010.

Stevens, Laura, Serena Ng, and Shelly Banjo, "Behind UPS's Christmas Eve Snafu," *Wall Street Journal*, December 26, 2013. As of November 14, 2016: http://www.wsj.com/articles/SB10001424052702303345104579282432991595484

University of Cambridge, "Ethnographic and Field Study Techniques," web page, undated. As of March 23, 2017: https://camtools.cam.ac.uk/wiki/site/e30faf26-bc0c-4533-acbc-cff4f9234e1b/ethnographic%20 and%20field%20study.html

U.S. Fire Administration, "Special Report: The After-Action Critique: Training Through Lessons Learned," Washington, D.C., USFA-TR-159, April 2008.

U.S. Fire Administration, "Operational Lessons Learned in Disaster Response," Emmitsburg, Md.: Federal Emergency Management Agency, June 2015. As of April 17, 2017:
https://www.usfa.fema.gov/downloads/pdf/publications/
operational_lessons_learned_in_disaster_response.pdf

U.S. Department of State, *Ensuring Leadership in a Dynamic World: Quadrennial Diplomacy and Development Review*, Washington, D.C., 2015. As of April 17, 2017:
https://www.state.gov/documents/organization/267396.pdf

U.S. Department of State, *Strategic Plan FY 2014–2017*, Washington, D.C., March 17, 2016. As of February 8, 2017:
http://www.state.gov/documents/organization/223997.pdf

vom Brocke, Jan, and Michael Rosemann, eds., *Handbook on Business Process Management 1: Introduction, Methods, and Information Systems*, 2nd ed., Berlin: Springer-Verlag, 2015.

White, Mark, and Alison Cohan, "A Guide to Capturing Lessons Learned," Nature Conservancy, undated. As of April 17, 2017:
https://www.conservationgateway.org/ConservationPlanning/partnering/cpc/Documents/
Capturing_Lessons_Learned_Final.pdf

Wiklund, George, and Lisa Graf, "Risk, Issues and Lessons Learned: Maximizing Risk Management in the DoD Ground Domain," briefing, Washington, D.C.: U.S. Department of the Army, October 2011. As of February 8, 2017:
http://www.dtic.mil/ndia/2011system/13235_GrafThursday.pdf